SECOND EDITION

SOILS FOR

LANDSCAPE

DEVELOPMENT

SELECTION, SPECIFICATION AND VALIDATION

SIMON LEAKE AND ELKE HAEGE

CABI is a trading name of CAB International

CABI
Nosworthy Way
Wallingford
Oxfordshire OX10 8DE
UK

CABI
200 Portland Street
Boston
MA 02114
USA

Tel: +44 (0)1491 832111
E-mail: info@cabi.org
Website: www.cabi.org

Tel: +1 (617)682-9015
E-mail: cabi-nao@cabi.org

A catalogue record for this book is available from the British Library, London, UK.

ISBN-13: 9781836990031 (paperback)
9781836990048 (ePDF)
9781836990055 (ePub)

DOI: 10.1079/9781836990055.0000

Edited by Joy Window
Cover design by Cath Pirret

Typeset by Envisage Information Technology
Printed and bound by CPI Group (UK) Ltd, Croydon, CR0 4YY

Foreword to first edition

In 1999, Tony McCormick of Hassell asked landscape architect Bruce Mackenzie and our office to take the design lead on what was to become the Sydney Olympic Park. Time was short and when we visited and reviewed the site, my heart fell. It was a desolate industrial wasteland that had been over time an abattoir, a munitions yard through two wars and, since World War II, a series of chemical plants had been dumping waste there for decades. Large parts of the 405 ha (the size of New York's Central Park), including Haslams Creek, could not sustain plant or animal life. The wasted soil was black and slimy. On top of this, the government required that none of the ruined soil could be removed from the site.

Tony had put together a superb team of consultants covering a wide range of ecological, botanical, and water specialists, as well as historical and political experts. Among these was the great Australian soil scientist, Simon Leake.

Step-by-step with Simon and the team, we fashioned a design that would deal with these formidable problems, reshaping the terrain into a series of hills, both symbolically geometric and naturalistic. Also included was the reconstruction of Haslams Creek and a park made up of major reforested tree walls, hills, and open meadows replanted with native grasses.

At the top of the site's major hill, the Olympic Center was designed by another architectural team with extensive paving and mature trees to accommodate the Olympic crowds, while our reforestation consisted mainly of tree seedlings and grass. Ten years later, the tree plantings have thrived, surpassing the size of the specimens in the Olympic Center, due, in large part, to the brilliant reconstitution of the soils under the 10-year direction of Simon – a miracle, in our experience.

Ten years later when we were awarded the commission to design the rebuilding of Barangaroo into a headland park, we again turned to Simon and landscape architect Stuart Pittendrigh to produce another miracle of soils and plants.

In this remarkable book, Simon and Elke Haege tell these and many more stories in scientific terms, extracting from them a series of principles and suggested procedures that every landscape architect, contractor and soil supplier needs to know to accurately specify and manufacture soils for landscape projects. Elke's synthesis of the world literature on recommended tree rooting volumes is a first.

Over the last 14 years, Simon has opened our eyes, and I strongly recommend this book to all professionals and administrators who are engaged in building and, particularly, rebuilding our sorely damaged urbanised world, in order to bring it to new life, utility, and beauty.

Peter Walker, FASLA Landscape Architect
Senior Partner, PWP Landscape Architecture

Contents

About the authors

Simon Leake and Elke Haege are the winners of the 2015 Australian Institute of Horticulture (AIH) Literature Award and winners of the 2014 Australian Institute of Landscape Architecture (AILA) NSW Research and Development Award for their 2014 *Soils for Landscape Development: Selection, Specification and Validation*, CSIRO Publishing.

Simon is a certified professional soil scientist with a particular interest in urban soil science. He runs a busy soil laboratory, SESL Australia.

Elke is a practising registered landscape architect, Fellow of the Australian Institute of Landscape Architecture, consulting arborist and horticulturist. She is passionate about sustainable and regenerative development of natural systems in urban environments.

Simon and Elke have created this book to assist, connect and elevate their related industries towards improved landscape outcomes.

Preface

Soils for Landscape Development: Selection, Specification and Validation is written to instruct and assist designers to specify landscape soils properly using objective and measurable criteria. It provides essential information and a universally applicable method for landscape architects and designers, specification writers, landscape contractors and soil supply companies using a systematic, clear and practical template based on solid and scientifically objective criteria.

This book links the landscape design processes with sound modern soil science practices to promote better quality project outcomes by ensuring that the basis of the landscape, the soil, is suitable for the intended purpose. Many different types of soils are used in landscape developments, from recycled site soils to lightweight podium and planter box mixes using both organic and artificial components. Many different proprietary soil products are available. This book attempts to provide an objective set of criteria for ensuring the chosen soil is truly suitable for the intended landscape treatment. The authors emphasise the use of measurable chemical and physical properties, well understood by soil scientists, to inform these decisions, not the use of arbitrary recipes or uncontrolled formulations.

This book seeks to provide landscape practitioners with a set of tools that will enable them to deliver functional and successful landscape soils to their clients and reduce the incidence of failure and the associated costs and wastes of natural resources. A strong emphasis in all chapters is placed on reducing environmental impacts by reuse of site soil, promoting appropriate minimal soil intervention and using recycled products.

Unique features of the handbook include:

1. *the Soil Design Approach Method*, which promotes consideration of existing site resources and outlines steps to take to minimise costs through soil reuse and to use site soil analysis and edaphic considerations to inform and promote sustainable soil design
2. *a site soil characterisation and investigation specification*, which provides a method by which site-won soils should be characterised and then used to form part of the performance specification. Thus the performance specification process is not confined to the use of purchased imported soils.
3. *performance-based technical specifications*, where two compulsory specifications are given outlining the performance requirement and the range of physical and chemical properties essential for the performance criteria to be met
4. *a performance description* where a clear description of both the intended use and type of soil mixes that might meet the criteria is given. This informative descriptive element allows for variation and market competition.
5. *suggested soil components*, which provide examples and guidance to soil supply companies as to the likely proportions in order to meet the specification criteria
6. *eleven main and three specialist landscape soil product specifications in a 'copy and paste' template*, which is ready for use in landscape soil specification reports. These

are numbered so they can be universally referenced. The most typical landscape soil situations have been rationalised to fit into these 14 categories.

7. *a set of design drivers* to improve the adaptation of landscape to place, and hence success rates, and sustainability, and provide optimum landscape performance

8. *a ready reference to landscape soils* to inform and promote sound reasoning for client understanding and satisfaction.

9. *three landscape soil verification and quality assurance specifications* with steps and guidelines to enable clear, standard, industry-accepted processes over the quality of the product, certification requirements, construction process, and correct supply of specified landscape soils and mulches

10. *the soil volume simulator* and outline of *eight key factors* in determining soil volume for trees in spaces where soil is limited.

The authors (Simon Leake and Elke Haege) have been advocating the use of objective measurable soil properties to specify soils and ensure they meet the performance criteria required, in preference to the current 'formulation-based' method so widely used in landscape construction contracts.

Since the publication of the first edition of this book in 2014, we have improved and updated the set of the most typical soil specifications that can simply be used by landscape development professionals for the majority of their landscape projects.

Our aim was to provide a clear and defined set of core specifications for the selection, specification and validation of soils for landscape developments that can be directly used for the majority of landscape projects in order to optimise the quality and longevity, and hence sustainability, of landscape design concepts.

Our purpose is to permit proper measurement-based quality assurance during construction to reduce the likelihood of landscape failure due to the purchase of soil that is inappropriate for the intended use and often not necessary. We also provide general guidance to soil selection and installation, with examples of the typical soil specifications. The general aim is to avoid the common plant and installation failures that are unnecessarily common in our industry.

We believe it is important to give landscape development professionals sound reasons why soil selection, specification and validation are important, and have therefore provided some case studies and some general background in a practical and understandable, yet scientifically sound, format.

We hope the use of this handbook will help to rationalise the varied array of chemical and physical property values currently provided to soil suppliers and contractors, so that the suppliers of soil will also use these specifications to develop soil products that meet the typical soil examples given. In this way, the process should become more streamlined and the industry self-regulated.

The utilisation of on-site soils in urban, suburban, pastoral or any development situation is common and the specifications encourage and allow for assessment, preparation, improvement and validation of site soil, as well as imported commercial soil.

The soil specifications are typically used to construct both the topsoil and the subsoil depths necessary for acceptable to optimum plant or tree growth. In the case of trees with limited space, additional advisory specifications are provided to define soil volumes (depth and extent) for tree growth zones to quantify and validate estimations for the area of soil required for adequate supply of nutrients, water, air, physical stability and hence growth and longevity for a tree or group of trees in urban environments.

The authors would like to thank the soil suppliers and the many landscape industry leaders and landscape contractors for adopting and specifying the soil performance specifications both locally and internationally over the 10 years since the first edition of this book.

We can see the improvements in not only the performance and outcomes of landscapes over the last 10 years, but also in the validity and increased importance of the professions, and furthered knowledge and collaboration gained between soil suppliers, waste recovery companies, soil scientists, landscape architects and designers, and landscape contractors to the betterment of the urban environment.

The authors would like to thank their associated industries for the support, and at times 'demand', for this handbook as being 'desperately needed' to iron out many real issues faced by practitioners, contractors, companies and developers, as well as the project outcomes as a result of these industries' combined efforts.

Of note, the authors would like to acknowledge the assistance of Chantal Milner and Owen Guy of SESL Australia.

Thank you to those who we consider the 'father figures' of soil-related matters in Australia: Kevin Handreck and Peter May. In addition, the authors thank the following leaders in their disciplines: former Land and Environment Commissioner NSW Judy Fakes and the Barangaroo Development Authority (BDA) for providing great knowledge, guidance and experience; Bruce Mackenzie; Graham Fletcher, UNSW; Pam Hazelton, University of Technology Sydney, and Dr Greg Moore, University of Melbourne, Resource Management and Geography and inaugural president of the International Society of Arboriculture, Australia Chapter.

Thank you to professionals of the landscape architecture industry including Adrian Pilton and Adam Robilliard of JPW, Mark Blanche of AECOM, Turf Design Studio, Aspect/ Oculus, and Guy Sturt.

Also, thank you to soil industry professionals Rob Niccol of Australian Native Landscapes and Tony Emery of SoilCo for useful suggestions on practical aspects of soil production. A special thank you to Sue Barnsley of Sue Barnsley Design for pioneering the use of performance specifications and quality assurance in landscape design and construction documentation. Also, to Peter Walker of Peter Walker Partners California for his trust in our approach on major projects in Australia.

1

Introduction

1.1 WHY ARE THESE SOIL SPECIFICATIONS NEEDED?

A definition of landscape soil

Landscape soil is an anthropic soil profile that is either modified from a natural *in situ* soil or manufactured and installed using artificial components for the purpose of sustaining vegetation chosen for landscape design or land rehabilitation.

As traditionally sourced landscape soil resources are running out or are no longer allowed to be mined or dredged from natural sites, the way we as an industry specify and install soils is and has dramatically changed.

In the first edition of *Soils for Landscape Development*, we introduced two pivotal changes in how landscape soils are designed and specified being:

1. the Soil Design Approach Method (see Chapter 4 'Soil design'), where priority is for the reuse and amelioration of existing site soils
2. the soil performance specifications (refer to Chapter 6), which pivoted traditional soil specification from a failed 'recipe-based' approach to a scientific, performance-based approach, specifying both physical and chemical properties of 14 typical soil landscape types.

This second edition streamlines the Soil Design Approach Method into a hierarchy including consideration of carbon emissions, sustainability practices and environmental performances that can be utilised in approaches to rating and measurement tools such as 'Towards Absolute Zero', 'Transform to Net Zero', 'GHG Protocol', 'Scope 1, 2, and 3 Carbon Emissions', 'Biodiversity Design Guides' and 'Green Star Rating', 'National Australian Built Environment Rating System' (NABHERS) and 'International Sustainability Rating System' (ISRS) performance rating systems.

This second edition incorporates an additional two example performance specifications:

1. Specification E4: Ultra lightweight growing media 'A' only horizon (for extensive rooftops with shallow growing media profiles)
2. updates to Specification F2: Raingardens, biofiltration and stormwater filtration soils.

This second edition also incorporates a section for small or lower budget projects and a simple to follow step-by-step guide on how and when to sample, test and stockpile, what to ask the soil suppler and how to order, variations to the specification and how to sign off. This section is great for low budget projects as well as for landscapers installing or maintaining residential or smaller gardens or builders, arborists, turf carers and home gardeners.

The industry has shifted to rely more heavily on soils that include waste recovery materials and recycled soils. It is therefore paramount that a performance based approach is adopted across the industry to ensure proper compliance. It is heartening to see, since this book's first edition in 2014, the industry widely adopting the performance-based approach and recognising soils as an important and valuable site asset and a fundamental step towards sustainability.

We have recently seen in the media whole landscape project sites shut down due to contaminated soil and/or mulch products. While this may not always fall to the responsibility of the landscape designer, these highly publicised news stories certainly reflect on the industry and can tarnish the reputation of the whole project team, the industry, and in themselves should be a prompt to ensure that correct validation and compliance statements are obtained.

We urge landscape architects and designers to take the project lead in proper soil design in all phases of the project process. On larger projects this would include engagement of specialist soil technologists or soil scientists to provide expert advice.

1.2 THE FUNCTION OF THIS HANDBOOK

When we first published *Soils for Landscape Development* in 2014, the industry in Australia relied upon a combination of an in-house 'recipe-based' specification, Australian Standard AS 4419 and NATSPEC to develop specifications for construction tenders.

These approaches were too broad and non-specific to provide any precision to the many design situations faced by normal landscape architecture practices, often resulting in the supply of soils that are not fit for purpose. Soil companies vary in their level of product quality and fitness for purpose, seldom test every batch and are also influenced by the economics of materials supply.

None of these approaches used performance- or technical product-based specifications in the same way that, for example, the concrete industry does to ensure the product is always fit for purpose and meets certain criteria. Some soil suppliers, then and still now, provide what might be called performance specifications in the form of product information sheets, but these vary in both the detail and accuracy of the representations made on such information sheets.

This industry-wide problem occurs across many commercial suppliers both in Australia and overseas and has been acknowledged in overseas-based reference books and at international workshops on the subject.

Soils for Landscape Development provides an objective, science-based approach to choosing and specifying soil media for landscape design, instead of providing vague 'recipe-based' specifications that often rely upon brand or proprietary products from manufacturers. It sets out key chemical and physical properties the soils must meet, as well as providing information about how this can be achieved. This approach is unique in the literature on landscape soil specification.

1.3 WHO WILL USE THIS HANDBOOK?

This handbook is written for landscape architects and designers, landscape construction contractors and soil supply companies. The landscape industry has used the first edition of *Soils for Landscape Development* to better understand the soil resources they may have on a site or to better understand how to specify soils using the 'copy and paste' individual specifications. Landscape contractors are using this book to ensure soil suppliers provide them with soils and soil ameliorants suitable for the project.

Soil supply companies are using this book to be informed of the typical soil specifications that landscape architects and designers are likely to specify, and positioning their products to be pre-compliant by meeting the specifications with independent soil analysis and 'fit-for-purpose' certification to their product batches.

Soil supply companies needing pre-compliant 'off-the-shelf' products that meet the specifications can quickly provide up-to-date compliance certificates and hence avoid the cost and time impositions on each project. This also provides an easier, lower risk, and more cost-effective approach, particularly for smaller scale projects. The pre-compliant products also typically provide a suitable base medium for minor modifications to meet large scale specialist project requirements.

The quality control specifications (G1-G3) provide the mechanism for certification via properly constructed 'product representation sheets' (i.e. pre-approved soils meeting the specifications). Also, there is room for validation of minor variations where local supplies mean the specifications cannot be met strictly, provided a professional expert is giving the opinion.

1.4 KEY BENEFITS

We hope that the updated and added information and guidance in this second edition, together with the updated 'example product specification templates' presented in this book will continue to move the industry from the traditional 'recipe-based approach' to a more sound urban soil science practice.

The specification method developed is universally applicable and it is able to provide an international standard in supply and installation of landscape soils regardless of the raw material at hand in any locality, although the specifications may need to be adapted to local conditions and practices.

Landscape architects and designers will be provided with a valuable resource that they can directly apply to their specifications, which will, in turn, improve the quality of their project outcomes.

It will also continue to improve the industry standards and allow a more level and fair tendering process for landscape contractors and a more level platform for soil supply companies. It will also assist to rationalise the number of soil products and product variations available.

This sound research and clear layout provides an easy-to-use, systematic compliance process that is internationally applicable and will contribute to ensuring landscape soil meets the specifications before installation.

1.5 STRUCTURE OF THIS HANDBOOK

This handbook is structured in a step-wise manner according to how a typical project development and landscape design process occurs. These steps include:

1. setting the project objectives
2. directing and conducting site analysis and soil investigation
3. generating informed design concepts
4. providing detailed design with specifications
5. enabling sound construction supervision.

Table 1.1. Steps in a typical project process

Steps	Details
Determine the project objectives.	E.g. budget, scope, aesthetics, function, maintenance and other requirements.
Analyse and characterise the existing site soil.	Include depths, type, structure and other properties. Basic or full soil analysis.
Determine the Soil Design Approach Method.	Prioritise for the reuse of site soil or the amelioration/remediation of the site soil. Can site soil be protected *in situ* or can it be stockpiled? Are there some parts of a project that might need imported topsoil/subsoil/subgrade?
Use the three steps above to inform the design.	Soil type and properties may influence plant choice and design.
Document the stockpiling and stripping.	Include the Soil Design Approach Method to be noted on the design tender and documentation drawings.
Specify the soil profile depths and number of horizons for reconstructed soils.	Such as an A/B soil profile or and A/B/C soil profile (refer to Sections 4.3 and 4.4 on creating soil profiles), e.g. horizon A: topsoil, horizon B: subsoil.
Choose the individual soil specifications required.	Include in the soil specification types on the design tender and documentation drawings.
Validate the product specifications in the documentation.	Outline in the specification the appropriate method and detail required for product certification and ordering, signing off on landscape soil testing and approvals of site soil works installation.

1.6 HOW TO USE THIS HANDBOOK TO DRIVE YOUR PROJECT

Table 1.1 outlines the series of steps in a typical project process. In order to create the most suitable performance specification, it is important to follow this process.

1.7 DUE DILIGENCE AND RESPONSIBILITIES

It is important to acknowledge and take into account that each site has varied soil characteristics and each project different objectives.

The example landscape product specifications in this handbook are ready to use templates and are based on typical landscape applications and soil needs for common landscape projects. They do not provide a complete list of all possible soils used in landscape projects and in many cases expert soil science advice will be needed to modify, vary or produce new specialist specifications.

The authors have considered the commercial practicalities of providing a core set of landscape soil products, as well as the promotion of recycled and waste material reuse so that the industry may be able to meet a more even product standard that still provides good to optimal plant growing conditions.

It is the responsibility of appropriately qualified and licensed landscape architects, horticulturists, arborists, landscape designers and landscape contractors involved in the project to ensure that appropriate and suitable analysis, investigation, characterisation, application and certification of these specifications are completed in response to the site and project conditions.

Further, methods of soil analysis vary both nationally and internationally. The suggested methods quoted in the specifications here have been used by the authors for many years, but may not be universally available or applicable. Where this is the case, the performance specifications should be used as a template and expert local advice from agronomists and soil scientists on what other acceptable methods and desirable ranges pertain should be sought and applied. Further, should a project site or design aspects have unusual, specialist or unique soil characteristics that are not covered in the standard specification templates presented in this handbook, it is recommended an agronomist or other suitably qualified soil scientist be engaged early in the project initiation phase to develop these specialist applications for tender using a method similar to that advocated here.

This problem of alternative test methodologies or specialist applications does not invalidate the method or importance of setting performance standards or specifications and, with time, a practising landscape architect or firm will develop the range of soil specifications and performance specifications adapted to local materials and soil-testing practices.

Although the creation of these specifications is based on sound scientific data and experience, the authors can take no responsibility for the use or misuse of these specifications, nor the variability of soils.

If any doubt over compliance occurs, employ the services of expert soil scientists to provide 'fit-for-purpose' compliance certificates or statements.

Introduction to soil fundamentals for landscape architects and designers

2.1 SOIL FUNDAMENTALS SUMMARY – FOR LANDSCAPE ARCHITECTS AND DESIGNERS

Landscape architecture as a profession has to encompass a broad combination of design, science, ecology, culture and community and coordinate with the construction industry.

Soils and soil design approaches are not only 'living systems' in themselves, but are also an absolute fundamental basis and an essential requirement for any successful and sustainable landscape.

Further, any landscape design that attempts to be truly sustainable must start with a soil design approach that will support the health, long-term sustainability, and commonly the regeneration of that living landscape. (Refer to Chapter 4 for the Soil Design Approach Method.)

Proper soil design approaches can also address soil contamination, which is a restorative landscape action and part of designing 'regenerative landscape systems' and designing 'intrinsic sustainability' into projects.

This chapter outlines the basic principles of soil science to allow a fundamental understanding of what constitutes a healthy soil and to recognise faults and diagnose problems or know when to seek specialist help. It will also allow a better understanding of the specifications and the properties they are trying to control.

The teaching of soil science usually divides the science into four fundamental subjects:

1. soil profile (pedology)
2. soil physics
3. soil chemistry
4. soil biology.

Before we do that however, we must introduce the concept of 'state factors.' Introduced by Hans Jenny in 1941 the state factor formula is a conceptual model that examines the different factors that influence soil formation. It says the soil properties in any given location are a function of the geology, climate, living organisms, topography and time:

$$\text{Soil} - \text{fn}(p, cl, o, r \dots t)$$

where:
* cl = climate
* o = living organisms
* r = relief or topographic position
* p = parent material or geology
* t = time.

'State factors' explain many things about soil formation or 'pedogenesis'. For example, in high rainfall climates basalt forms a bright red/brown clay loam soil but in drier climates a dark black cracking clay soil. It explains why very new soils such as those in a hind dune at the beach show almost no obvious horizons but a very old podsol soil on such sands form several striking horizons such as a bright white A2 horizon and a bright orangey brown B horizon.

Soil is essentially made by living organisms, particularly plants that concentrate nutrients and add organic matter to the all-important topsoil in a kind of self-fulfilling prophesy resulting in better conditions for plant growth.

Essentially, organisms, time and climate combine to form the soil profile.

2.2 PEDOLOGY – THE NATURE OF SOIL PROFILES

When excavated and exposed soils most commonly show a 'profile' composed of different layers or 'horizons' (refer to Table 2.1). We call this 'differentiation'.

The older the soil, the greater the differentiation and the clearer the horizons.

This represents a well-developed soil ~1.2–1.8 m deep on fine sediments like shale but similar profiles will also form on sandstone and volcanics like granite. Igneous rock (fine-grained rocks like basalt) do not tend to form the pale A2 horizon. Very young soils and soils in arid environments don't show such differentiation into horizons, very young soils simply because there has not been enough time and arid soils because living organism activity and leaching by rainfall is limited.

Some important rules for the construction of landscape soils come from this profile analysis:

- Organic matter is only ever in the surface, ~300 mm but mostly < 200 mm deep. Organic matter is never found at depth.
- Texture always gets heavier (higher in silt and clay fines) with depth, never the other way around. Therefore, when constructing soil profiles, never put clay soil on top of sand.
- A well-functioning soil will have as a minimum three horizons: A, B and C.

O Horizon (Organic matter)
Mulch layer of decomposing organic matter

A Horizon (Topsoil)
Higher organic matter, lighter textured, friable soil with appreciable nutrient content.

B Horizon (Subsoil)
low organic matter, typically heavier textured soil with some nutrient content.

C Horizon (Subgrade / parent material)
Very slightly weathered rock (parent material).

Fig. 2.1. Graphic and photo of a soil profile illustrating the different soil horizons.

Table 2.1. The most common characteristics of the horizons

Horizon name	Description	Function	Constructed landscape analogy
A0	Partially decomposed organic matter at the surface. Always of lighter (less clay) texture than the subsoil.	Adds organic matter and nutrients as it decomposes. Moisture conservation	Mulch
A1	Soil layer with the highest organic matter and nutrient content. Paler than A1, sometimes almost white.	The root zone. Zone of highest root colonisation. Moisture and main nutrient store.	Topsoil
A2	Topsoil with a lower organic matter and nutrient content.	Root anchorage, moisture and nutrient store.	Topsoil
B1	Subsoil, always of heavier texture (higher clay content) than topsoil above. Always the zone of brightest colour, e.g. yellow and red brown.	Root anchorage, moisture store. Little contribution to nutrition.	Subsoil
B2	Deep subsoil. Duller colours than B1 and less clay.	Drought moisture store.	Subgrade
C	Gravelly decomposed rock.	Weathers to form more soil.	Subgrade or basement rock

Describing soil profiles gives rise to the science of pedology, which includes the classification of soils into similar groups.

Soil classification systems differ nationally and internationally. Soil description and classification systems commonly used in Australia are found in Northcote (1979), Stace *et al.* (1972) and Isbell (1996) In the United States of America, Craul (1992, 1999) provides detailed information on assessing field soils for landscape work.

Soil classification is not really required by the landscape development industry but for the purpose of completing formal soil reports the soil profile types can be found in public soil surveys and now online at such websites as eSPADE v2 (https:/www.environment.nsw.gov.au/eSpade2Webapp/) for New South Wales, Australia.

It is important to be able to recognise soil horizons for such purposes as establishing stripping depths for topsoil and subsoil and rebuilding soil profiles from recovered soil horizons.

2.3 SOIL PHYSICS

Soils are composed of solids, air and water in various ratios. A healthy soil must have air because roots, fungi and microorganisms are living, respire and so consume oxygen. Living organisms in the soil, including plant roots, must have water and the dissolved minerals and nutrients within are transferred through uptake of water.

Mineral and solid components also provide physical support. The water and air fit into the pore space between the solids for the 'living system' to use.

The solid fraction is composed of various components:

- *Mineral solids:* sand, silt, clay, typically from weathered substrate rock or deposited sediments
- *Mineral nutrients:* macronutrients and micronutrients that are essential for plant growth are all soluble mineral elements in the form of 'ions' (see Section 2.6 below).

- *Organic matter:* this includes decomposing plant matter which becomes dark 'humic' matter over time (with the aid of the living organisms, air and water and time). Organic matter holds moisture, assists in soil structure, adds nutrients and eventually all converts back to carbon dioxide. It is the energy source or food for soil organisms.
  - ▶ Carbon, or soil organic carbon (SOC), is created and stored in this zone.
- *Soil organisms:* insects, worms, bacteria, fungi, algae, and many other types of microorganisms which can metabolise organic matter to release nutrients to make them available for plant uptake again.

The solid fraction of soils should ideally show ~25–40% pore space between them. This results from the soil having a structure whereby the soil particles join together into units called 'peds' with micropores within the ped and macropores between the peds. Organic matter together with microorganism activity acts as the 'glue' that sticks the peds together but clay minerals also can form well-structured soils.

A poorly structured soil will be dense with little pore space, poor water permeability and poor water holding ability and this results in poor root and plant growth. Structure can be destroyed by excessive ploughing, compaction by vehicles and pedestrian traffic and allowing the organic matter content to run down.

Permeability is the ability of a soil to conduct water: its 'hydraulic conductivity'. It is important for soils to accept water from rainfall and irrigation. The two main factors that lead to poor permeability are:

1. a high content of poorly structured silt and clay
2. compaction.

Fig. 2.2. Image of an excavated lawn showing a uniform and highly permeable (sandy) soil with poor pedality resulting in poor water holding capacity. Soil nutrients for plant growth will readily leach through this soil.

Fig. 2.3.   This image of an exposed soil profile on a construction site shows compacted fill under a pavement. This soil profile is compacted with a high clay content and much lower permeability (and poorer drainage abilities) than the sandy soil profile in Fig. 2.2. This soil profile can be prone to waterlogging if the water is not redirected away at the pavement surface, but can also be effective at retaining soil moisture, stability and soil nutrients. It can also be prone to higher amounts of swelling and shrinkage with irregular rain or irrigation.

Fig. 2.4.   This soil profile shows a developed, weathered roadside cutting demonstrating well-formed peds and soil structure (aggregation). This profile with well-formed peds and clay content provides good water holding ability. Compare this soil profile with the compacted soil profile in Fig. 2.3.

This is why the turf soil and sportsfield soil specifications place strict upper limits on silt and clay content and lower limits on the sand fractions, because they receive high levels of pedestrian traffic which causes compaction.

2.4 UNDERSTANDING COMPACTION

Compaction is caused by pressure on the ground surface. That could be a truck tyre or a human foot. The pressure exerted by a tyred vehicle is exactly the same as its tyre pressure. So if a passenger vehicle has tyre pressure of 25 psi, that pressure is applied to the surface. A truck on the other hand has tyre pressures of 120–140 psi. A footballer

Fig. 2.5. Image of a sports field with compacted soil due to poor soil particle size distribution and not 'fit for purpose' for the intended use.

running on the balls of their feet can exert up to 50 psi. And that's not counting the pressure the sprigs apply.

Soil moisture also affects compaction. Soil strength declines as soil becomes wet. The common practice of holding farmers' markets in parks where trucks and vans are taken into the park, even if the soil is wet, can cause significant damage to soils resulting in decline of the turf and park trees, and decreased infiltration of water leading to drought-like conditions to plant roots. This can be irreversible, or at least expensive to repair.

The main physical testing required to understand how to achieve the soil design for the modification of the soil for any given purpose includes:

- *Particle size distribution (PSD):* where compaction resistance and permeability are critical, such as in sportsfield and biofiltration soils, specifying soil texture alone is not enough information and a full PSD is required.
- *Texture:* Soil permeability is directly influenced by the amounts of sand, silt and clay (the texture), and soil air pockets, so soil texture analysis is a very informative test.
- *Structure:* This also relates to the permeability of the soil and soil porosity. Refer to Figs 2.3–2.5).
- *Permeability:* Where this is critical in sportsfield and biofiltration soils, permeability should be measured. This is best measured *in situ* or by using steel cores to take intact samples. Conducting permeability testing on disturbed samples is often not representative of the real behaviour in the landscape. It is only suitable to conduct 'repacked' permeability testing on very sandy soils such as sportsfield and biofiltration soils. Refer Fig. 2.5.
- *Organic matter:* This informs soil amendments and maintenance requirements. The most suitable amount of organic matter for topsoils is typically between 3–10% by weight (however refer to the soil specifications in Chapter 6 for the range recommended for various different landscape applications). Note: additional organic matter can be easily added in the form of compost which is now cheap and readily available in all urban centres.
- *Wettability:* This is a property whereby the soil absorbs water rapidly, the opposite to hydrophobia where the soil repels water and will not wet up. It can be measured by various methods (see AS 4419 Appendix C). It is usually caused by the presence of very old organic matter which is dominated by waxes and resins that resist decomposition. It is particularly common in organic potting media. It is overcome by incorporating fresh organic matter and/or the use of soil wetting agents (see Section 2.5).

2.5 WATERLOGGING AND LACK OF OXYGEN

Strictly called 'soil hypoxia' (soil with limited oxygen), in its worst case, where oxygen is completely absent, it is called 'anaerobia' (no oxygen), these conditions, in our experience, kill more new landscape plants than drought. As stated above, roots respire and need both oxygen to diffuse towards their roots and carbon dioxide produced by respiration to diffuse away. Otherwise they suffocate and die. Refer to Fig. 2.6.

The things that contribute to hypoxia are:

- *Wet soil:* diffusion of air through water is 10 000 times slower than through air so if the pore space is filled with water oxygen simply can't diffuse towards the roots fast enough.
- *Buried organic matter:* the microbes, worms and insects that decompose organic matter all need oxygen to respire. If organic matter is buried too deeply and then the soil gets wet, water takes up air pockets, and it can rapidly become hypoxic, then anaerobic.

Fig. 2.6. Waterlogging leading to anaerobic conditions and death of new tree planting, due to poor subsoil drainage and over-watering. (Source: with thanks to Sue Barnsley Design.)

- *Compaction:* This means low pore space. If air pores become water filled the soil zone very rapidly becomes hypoxic.
- *Roots buried at depth:* In landscape construction, often containerised advanced trees are installed with container heights around 800 mm. Installing advanced plant specimens into soil that has low soil oxygen levels (which is typical below 300 mm depth), and or poor drainage from 300–500 mm depth, the root zone can very possibly suffer and/or succumb to low soil oxygen levels to the point of hypoxia. Installing aeration tubes can be a simple and cheap precaution. Trees naturally develop 90% of their roots in a shallow plate formation in the top 300 mm of soil (where there is more air, pore and organic matter available).

2.6 SOIL CHEMISTRY

The major constituents of plants are carbon (C), oxygen (O) and hydrogen (H). C is supplied as carbon dioxide from the atmosphere and O and H from water in the soil (water comes from the atmosphere as well). All the other nutrients come dissolved in water from the soil.

Plants take up mineral nutrients as dissolved 'ions' in the soil water. It is necessary to understand what an ion is. It's all about salt chemistry. A salt (sodium chloride or common table salt is just one type of salt) is something that has two parts, a positive and a negative, that when dry forms a solid, but when it meets water, it dissolves, and it does this by separating into two parts:

$$AB \rightarrow A^+ + B^-$$

They separate into one positively charged part (the A^+ or cation) and the negatively charged part (the B^- or anion). An example is common table salt:

$$NaCl \rightarrow Na^+ + Cl^-$$

Table 2.2.   The important plant nutrients and their ionic forms

Cations	Anions
Nitrogen: NH_4^+	Nitrogen: NO_3^-
Potassium: K^+	Phosphorus: PO_4^{3-}
Calcium: Ca^{2+}	Sulphur: SO_4^{2-}
Magnesium: Mg^{2+}	Boron: BO_3^{3-}
Sodium: Na^+	Molybdenum MoO_4^{2-}
Iron: Fe^{3+}, Fe^{2+}	Chloride Cl^-
Manganese: Mn^+, Mn^{2+}	
Zinc: Zn^{2+}	
Copper: Cu^{2+}	

Sodium (Na^+) is the cation and chloride (Cl^-) the anion. This is usually the case where the metal forms the cation and the non-metal the anion. (Refer to Table 2.2 for a list of plant nutrients and their ionic forms.)

It is important to understand that plants cannot take up these nutrients as complex organic food like we can; the organic matter must be decomposed and the nutrients 'mineralised' before the roots can take them up. Even then, there must be water in the soil for plants to take up the soluble nutrient ions. The saying 'water is the best fertiliser' is more than half true for this reason.

2.7 THE IMPORTANCE OF pH

pH is an assessment of the acid/alkali balance of the soil, and the pH scale ranges from 0 for highly acidic to 14 for highly alkaline with 7 in the middle being neutral. Generally, a landscape soil pH should be between 5.4 and 6.8. An out-of-range pH can indicate unsuitable micronutrient levels. Refer to Fig. 2.11.

pH determines the 'availability' or the solubility of certain trace elements, particularly iron and manganese but also zinc and copper. At alkaline pH these form very insoluble compounds, greatly decreasing their solubility and therefore availability to most plants. Only special plants adapted to alkaline (chalk or limestone) soils can get enough iron in such soils. These used to be called 'calcicoles' (as opposed to calcifuges which grow on acid soils only) but now we prefer the term 'iron efficient' plants. Plants that are not adapted to limey soils but only to acid soils are called 'iron inefficient' plants. This relates to the other metal trace elements like manganese as well. Clearly the *Pomaderris* in Fig. 2.7 is an iron inefficient plant showing extreme manganese deficiency when grown on alkaline soil. The species is adapted to acidic rainforest soils. (Refer to Section 3.16, 'Lessons from failed projects (three examples of common landscape soil failures)' for an explanation of how this deficiency arose.)

In acid soils other problems can occur:

- Aluminium, a common constituent of all soils, can dissolve and become toxic to plants.
- Manganese likewise can dissolve to toxic levels.
- Phosphorus is bound to the soil or 'fixed' and becomes unavailable.

Another example of a leaf showing potassium deficiency is in Fig. 2.8, showing a hydrangea leaf with yellowing leaf margins on older leaves. The comparison of this visible symptom with the magnolia leaf in Fig. 2.9 demonstrates that visual leaf

Fig. 2.7.   Severe lime-induced manganese deficiency in *Pomaderris ferruginea*. Refer to Section 3.16 for the full story about this 'failed landscape' and tips to avoid it.

appearance does not necessarily indicate definitive specific deficiencies, but it does encourage the experienced practitioner to obtain leaf analysis from a laboratory.

Soil analysis within the tree pit with the *Magnolia grandiflora* (Fig. 2.9) shows that the pH was moderately acidic. The organic matter content was low at 1.5%. The effective

Fig. 2.8. Potassium deficiency in *Hydrangea* typically shows yellowing leaf margins in older leaves. Potassium deficiency is commonly associated with older garden bed soils.

cation exchange capacity (ECEC) was close to being balanced, but magnesium and potassium were low. It was recommended to boost potassium with 40 g/m^2 of muriate of potash and add 20 L/m.

Soil analysis within the tree pit with the *Podocarpos elatus* (brown pine) (Fig. 2.10) shows the pH is strongly acidic with a low effective cation exchange capacity (ECEC),

Fig. 2.9.   Leaf analysis results of a *Magnolia grandiflora* that is very stunted and growing in a contained street tree planting hole shows multiple deficiencies of nitrogen, magnesium, calcium and potassium. Manganese was also very deficient, with high levels of copper and normal levels of boron.

Fig. 2.10.   (Left) Leaf analysis results of *Podocarpus elatus* (brown pine/plum pine) show deficiency in the macronutrients: nitrogen, magnesium, calcium and potassium. Iron is very high, and boron is deficient as is manganese. This tree is very stunted and growing in a contained street tree planting hole. (Right) This *Grevillea robusta* (Southern Silky Oak) had phosphorus toxicity that was induced by high available soil phosphorus.

indicating poor nutrient holding capacity. Most soil nutrients indicate nitrate, potassium, calcium and sulphate were low and required boosting. The organic matter content was high at 7.8%.

Given the small tree pit sizing, and already high organic matter content, the application of urea at 20 g/m^2 and sulphate of potash at 50 g/m^2 and lime at 200 g to help raise the pH levels and add calcium were recommended. Multiple deficiencies require detailed soil testing and carefully targeted nutrient enhancement.

The soil pH chart in Fig. 2.11 is useful in illustrating the reduction of each of the available nutrients (to varied amounts) as the soil pH becomes more acidic or more alkaline. This chart also shows the soil pH optimum 'neutral' range. A simple pH test kit is a useful and quick test to use when assessing soils on site. Refer to Fig. 2.12.

IDEAL SOIL pH RANGE

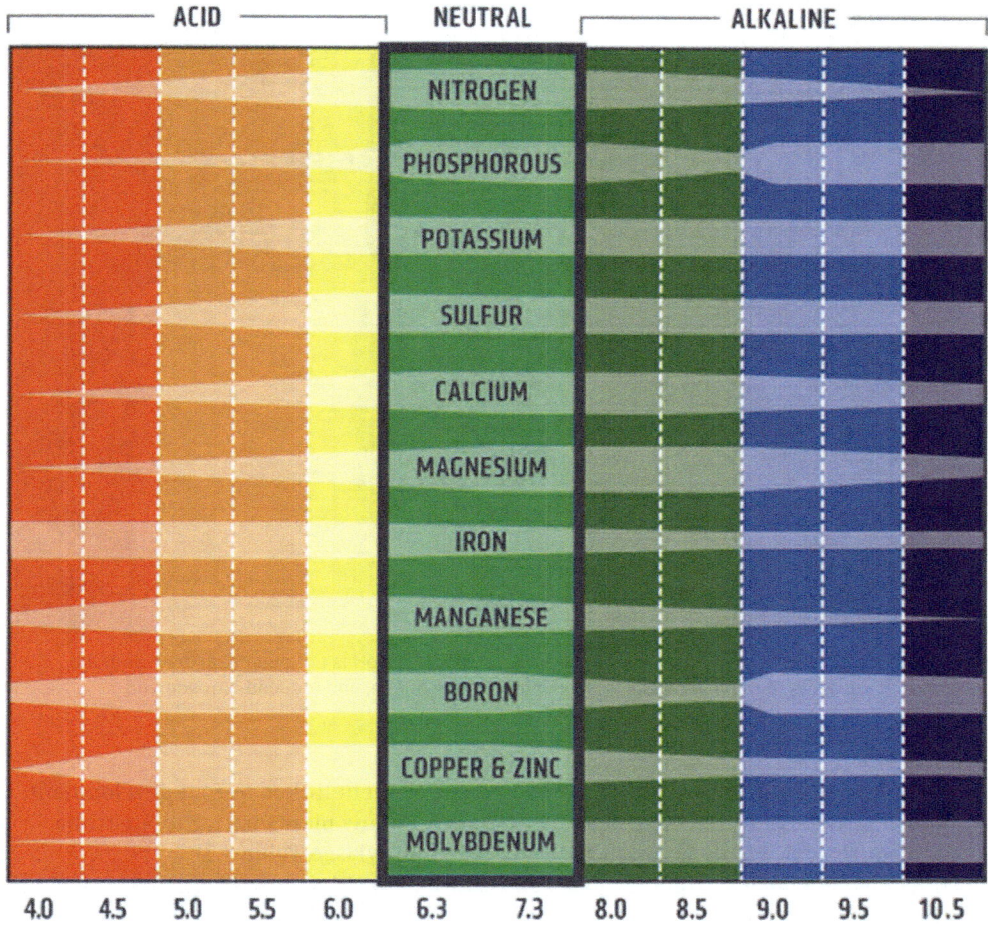

Fig. 2.11.   pH chart showing the availability of key nutrients at various pH levels. (Source: PROFILE Products LLC, <https://www.profileevs.com/products/hydraulic-mulch-additives/ph-modification>)

Acid soils can be easily rectified with lime but acidifying alkaline soils is much more difficult. To do this we use sulphate of iron or agricultural sulphur but if there is more than ~1% lime in the soil this becomes impractical. In this case the horticulturist must choose plants that are tolerant of alkaline soil such as many Mediterranean plants.

So the major chemical properties to take into account, and get tested when assessing site soil or quality assuring imported soils, are:

- *pH* as discussed above
- *electrical conductivity,* a measure of salinity of the soil, too much of which can cause damage to plants; where salinity is elevated it is useful to measure chlorine (Cl)
- *major cation nutrients* including sodium (Na), potassium (K) calcium (Ca) and magnesium (Mg)

Fig. 2.12.   A simple in-field pH test indicating an alkaline soil. This pH is not optimal and will limit the availability of nutrients to the plants such as iron, boron, manganese, and reduced nitrogen and phosphorus.

- *phosphorus*, a critical and most often deficient element linked to root development; phosphorus toxicity can also become an issue in many plants native to Australia and South Africa (particularly the Proteaceae family) (see Table 3.3)
- *trace element metals*, iron (Fe), manganese (Mn), zinc (Zn), copper (Cu)
- *the major anions:* nitrate (NO_3^-) and sulphate (SO_4^{2-})
- *the trace anion* borate (BO_3^{3-}).

Landscape architects and designers are encouraged from a sustainability position to value the existing site soil as a useable site asset, and a basis from which to design the Soil Approach Method, which will in turn influence the landscape design and ultimate success (or not) of the landscape outcome.

Often, however, soils in their native state are not fertile enough for the intended landscape design outcome the client desires, and the site soils need to be improved or ameliorated accordingly. Soil analysis will inform the designer as to the exact soil physical and chemical properties needed rather than just guessing or assuming.

Soil analysis can minimise the use of resources like fertilisers to only the necessary amounts and types, and avoids adding soil elements that are not needed. Soil analysis avoids costly rectification of failed or underperforming landscapes.

Some commonly added ameliorants and fertilisers are:

- lime and dolomite to correct acidic pH and add Ca and Mg if needed.
- sulphur to correct an alkaline pH

- gypsum (calcium sulphate) to stabilise clay soils
- compost to add organic matter and a wide range of balanced nutrients
- manures to add organic forms of nutrition mainly N and P.
- single element fertilisers to target a specific deficiency such as:
  - ▶ phosphate fertilisers to correct P deficiency
  - ▶ potassium sulphate or chloride to correct K deficiency
  - ▶ urea and ammonium sulphate mainly for the N content
  - ▶ trace element fertilisers mainly sulphates (e.g. iron sulphate, copper sulphate). These should never be used unless testing shows they are deficient.

Soil science is a complex profession requiring considerable experience with soil and plant analysis to diagnose problems. It is always better to get the soil specifications right in the first place and enforce adherence to them during construction than to try to correct problems afterward. Landscape architects and designers are encouraged to engage with competent soil scientists in their local region in forming their soil design approach (see Chapter 4) and specifications particularly when it comes to site soil improvement.

Designing the vegetation to suit specific soil attributes can reduce the amount and cost of soil amelioration needed, and is an important sustainability consideration that landscape architects, designers and horticulturists must consider as part of the soil design approach. Refer to Chapter 3, Table 3.3. for a list of commonly used plants that are

Fig. 2.13. Hydrophobic soil (water repellent) caused by dried out surface algae makes water sit on top or run off the surface and not soak into the plant's rooting zone. Action needs to be taken to improve the wettability of the soil by integration of manures (as suitable to the plants) and organic matter (as suitable to the plants) and regular irrigation frequency.

generally tolerant or sensitive to specific soil attributes (such as acid, alkaline, salt and phosphorus sensitivities).

2.8 SOIL BIOLOGY

Much is talked about the importance of soil biology and no competent soil scientist would deny it. Soil contains billions of living things of millions of different species and varieties including micro-forms such as bacteria, actinomycetes, fungi, protozoans, nematodes and macro-forms such as worms and insects.

All of these biological constituents are dependent on the cycle of organic matter turnover from plants to fuel their existence. Some have taken it a step further and actually invade plant roots to form a commensal or symbiotic relationship with the plant. Rhizobia bacteria, for example, can fix N from the atmosphere and give it to the plant as nitrate. In return the plant feeds the bacterium sugar and nutrient elements. Mycorrhizal fungi do likewise, connecting to plant roots and helping improve the availability of soil nutrients, especially phosphorus.

On the other hand, some bacteria, actinomycetes, fungi, protozoans, nematodes and macro-organisms such as insects can be detrimental to plants, like in any stressed, or out of balance ecosystem but also through the natural process of predation and decay.

Mycorrhizal fungi can be purchased to add to soils but be aware that these are highly host specific and may benefit only one or two or even none of the plants in your horticultural palette. Also, they have been found to only benefit plants in really poor, infertile soils, whereas by adding compost and some fertiliser the added fungal benefit greatly diminishes. Some monoculture forestry practices in the United States of America using a species specific to certain pines have been shown to benefit forestry production but, again, on very poor soils. Preparations of 'soil probiotic' products usually don't come with any objective experimental results and their claims should be treated with some caution until this specialised field develops and improves.

The soil organism ecosystem is very complex, but we have found that soils which have adequate, suitable and matured organic matter compost (generally between 3–6% that is actively breaking down and topped up periodically to stimulate the soil biota) will improve plant root growth, soil and plant health as it enlivens the soil biology. This is not the case for the hydrophobic soil shown in Fig. 2.13.

2.9 GENERAL FURTHER READING LIST

Handreck KH (2009) *Gardening Down-Under: A Guide to Healthier Soils and Plants.* 2nd edn. CSIRO Publishing, Melbourne.

Handreck KH, Black ND (2010) *Growing Media for Ornamental Plants and Turf.* 4th edn. UNSW Press, Sydney.

Hazelton P, Murphy B (2011) *Understanding Soils in Urban Environments.* CSIRO Publishing, Melbourne.

Jenny H (1941) *Factors of Soil Formation.* McGraw-Hill, New York, NY.

3

Understanding soils for landscape development

3.1 UNDERSTANDING AND INVESTIGATING SITE SOILS

The recovery and reuse of site soil resources during development should be the preferred option in every project where such resources are available.

Not only does the reuse of site soil represent sound environmental practice, reducing the need to mine soils from other locations, but very often a site soil resource is of better quality than commercially available soils. Site soils may, however, require amelioration or improvement to support the intended new landscape aims sustainably. Site soils may not be suitable for certain uses; for example, a heavy clay loam will not be suited to high-wear turf areas.

The purpose of the specifications (Chapter 6, Part I, **Specifications A1** and **A2**) is to prescribe a method to properly identify and characterise the on-site soil resources and develop and specify a stripping and reuse plan, properly enforce its implementation, and rebuild a profile with at least a minimum of functional purpose.

3.2 NATURAL SOILS

In the authors' experience it is not entirely essential to provide a precise soil classification for landscaping purposes. It is typically only necessary to provide sufficient information on the soil profile to instruct tenderers in reuse and recovery or any special planting methodology, such as if large trees are to be planted in a soil with an impermeable subsoil.

For practical purposes the primary profile forms described by Northcote (1979) provide a simple and easily understandable means of determining stripping depths and likely soil issues relevant to landscape developments. They are summarised in Table 3.1.

These five simple categories really provide all the information a landscape development soil surveyor needs to decide on sampling depths, stripping depths and likely management issues.

For example: If the soil profile is a uniform coarse sand, the stripping depth is not so important: just strip to provide enough topsoil for the landscape works. Also subsoil drainage will not be needed because the profiles are inherently well-drained throughout the profile.

If, however, the soil is a duplex or podsolic soil it is critical to strip the loamy topsoil cleanly and not contaminate it with clay subsoil. Further, special planting techniques and/or systematic drainage will be needed at specified depths when planting larger stock to prevent waterlogging of the root ball.

Table 3.1.   Primary profile forms (Source: Northcote 1979)

Primary profile form and principal features	Subdivision	Examples
1. Organic O: peaty organic matter at least in the topsoil, usually waterlogged or cold	None	Peaty soils of swamps and poorly drained areas
2. Uniform U: uniform texture throughout (e.g. sand in a	Coarse Uc	Coarse sandy soils of coastal dunes and sandstones
sandy soil or in beach dune areas)	Medium Um	Deep loams and sandy loams typical of desert or sandy coastal areas
3. Uniform U: uniform clay throughout the profile	Fine not cracking Uf Fine cracking Ug	Plastic clays that do not crack Cracking clay soils of the black soil plains
4. Gradational G: gradually becomes heavier (more clay) with depth	Calcareous throughout Gc	Calcareous red or black earth on limestone or marl.
	Not calcareous throughout Gn	Yellow earths common on coarse sandstone geology
5. Duplex D: have a distinct texture change from the A to the B horizon (e.g. loam topsoil to clay subsoil). Subsoil colour is important	Red Dr Brown Db Yellow Dy	Red Kurosol soils common on shale, better drained, top of slope Yellow Kurosol soils Yellow podsolics, usually lower slope
	Grey Dg	Grey clay subsoil, usually wet or hydrated

Engineering note: Engineers and earthmovers will often use the word 'soil' to refer to subsoil and broken rock material from deep layers. Ideally, this should be called 'fill' to differentiate it from soil, which is the surface covering material distinguished by the presence of organic materials and living organisms.

Soil information is often available from public institutions: most commonly departments of agriculture, planning or soil conservation services and environment protection authorities and agencies. The level of quality of publicly available soil surveys will vary nationally and regionally, but it always pays to do some desktop work first to obtain such site information as is available. Keep in mind that soils in urban areas are often highly disturbed and the soil survey information may be of only general assistance.

The bibliography provides references to soil mapping information available in Australia. With experience, and in simple soil landscape areas such as the very sandy areas of southern Sydney, Perth and south-east Melbourne, a landscape planner can often, through experience, provide an adequate understanding of the site soil resources for landscape development purposes from their own observations. In more challenging and complex areas, the services of a professional soil scientist are recommended.

On *disturbed sites*, the normal association of horizons can either be confused and mixed or will often not be present. Often the mixed soil/fill material present is so disturbed it is considered 'subgrade'. **Specification A2** provides a minimum suite of methods for providing information on whether the existing surface soil materials can be modified to form either topsoil or subsoil, or has to be relegated to deep 'subgrade'.

On disturbed sites, the services of a professional soil scientist or landscape technologist are strongly recommended to see if recoverable soil material is present and reusable.

Specifications A1 and **A2** represent the minimum suite of information on soil properties that should be collected to allow the following Soil Approach Method and processes to be defined:

- whether the soil material present can be used as topsoil, subsoil or should be relegated to subgrade
- how the soil conditions may provide opportunities or constraints on the design process
- topsoil stripping and recovery methods
- what adaptations or changes (improvement or amelioration) will be needed to the soil properties to use it in the landscape treatments
- what properties of the subsoil need to be considered in planning (e.g. salinity, sodicity, alkalinity)
- what amelioration the subsoil may require.

3.3 DISTURBED, ARTIFICIAL OR MANUFACTURED SOILS

Quite often the landscaper will be required to use the finished subgrade following construction as the subsoil for the project.

> Unless steps are taken during construction to save topsoil, then none will be available. Unless steps are taken to protect subgrade, the subgrade is likely to become disturbed, often mixed with and affected by site construction waste and unduly compacted or contaminated during the construction process.

Ideally, the soil stripping, stockpiling and recovery plan (**Specification B1**), if performed as part of the site analysis, will have anticipated the amelioration requirements of the subgrade to render it suitable as subsoil. This plan may have stipulated the recovery and replacement of some site subsoil or the processes by which the final site subgrade may be ameliorated.

Highly urbanised sites will frequently show altered, highly disturbed or completely destroyed soils. Isbell (1996) provides a useful way of describing such anthroposols, which are summarised in Table 3.2.

In reality, most soils in established city parks and urban areas are both cumulic and hortic, and it is common to find layers of sand applied to turf areas, as well as increased organic matter levels due to topdressing composts and mulches. It is also common to find foreign

Table 3.2.   Ways of describing anthroposols (Source: Isbell 1996)

Anthroposols	Soils resulting from human activities – either modification of pre-existing soils or creation of completely new soils
Cumulic	Soils which have formed due to humans depositing materials over time, such as shell middens or weathered rock cairns, or burial mounds
Hortic	Soils that have had organic materials, such as composts and manures, incorporated into the surface
Garbic	Newly created soils used to bury refuse landfills, usually high enough in organic matter to create methane or land fill gas
Urbic	Newly created soils covering fill of a predominantly mineral nature not creating methane. Most 'brownfields' sites
Dredgic	Soils forming on mineral materials dredged out of the sea or other waterways, including tailings ponds from quarrying and mining
Spolic	Soils formed on mineral materials disturbed by mining, road building and construction
Scalpic	Soils forming on areas where the upper soil layers have been removed or truncated

objects (glass, ceramic, metal and plastic) scattered through topsoil of urban soils. The definitions in Table 3.2 provide no information on the chemical properties likely to be encountered in any given location. The site investigation **Specifications A1** and **A2** provide a means of determining both the physical nature of the profile, as well as its chemical properties relevant to landscape development, and thus are applicable to intact as well as disturbed soils.

Designing the vegetation to suit specific soil attributes can reduce the amount and cost of soil amelioration needed. Table 3.3 lists some commonly used plants generally tolerant or sensitive to specific soil attributes (such as acid, alkaline, salt, and phosphorus sensitivities).

Further reading

Bassuk N, Curtis DF, Marranca BZ, Neal B (2009) *Recommended Urban Trees: Site Assessment and Tree Selection for Stress Tolerance*. Urban Horticulture Institute, Cornell University, Ithaca, NY.

Handreck K, Black H (2010) *Growing Media for Ornamental Plants and Turf*. 4th edn. NewSouth Publishing, Sydney. See pp. 525–531 for phosphorus-sensitive plants.

Hazleton M (2016) *Bartlett Tree Experts, Trees for Alkaline Soil*, <https://www.bartlett.com/resources/technical-reports/tree-species-for-alkaline-soils>, accessed June 2024.

The Australian Plants Society of South Australian Region, <https://www.australianplantssa.asn.au>, accessed January 2024.

3.3.1 Phosphorus-sensitive plants

Most Australian soils are naturally very low in available phosphorus for most exotic plants and, likewise, conventional garden soils have too much available phosphorus for some native plants; this will cause toxicity and probable failure of these plants if planted in the incorrect soil medium.

Table 3.3 lists commonly known phosphorus-sensitive plants thanks to research conducted by Kevin Handreck (published in Handreck and Black 2010).

When planting some Australian natives in new landscapes, it is important to keep the known P-sensitive plants in garden beds together with other phosphorus-sensitive plants and specify a low P soil mix. It is difficult, if not practically impossible, to mix them successfully with plants that require high P levels.

Phosphorus-sensitive species include many woody and herbaceous perennials in the Proteaceae, Mimosaceae, Leguminosae and Rutaceae families and some in the Myrtaceae (refer to Table 3.3). This group of P-sensitive species does not include all Australian native plant species, as is the commonly held misconception.

In Europe it is common to refer to plants that require acidic and low-fertility (including low phosphorus) soils as 'Ericaceous plants'. This is not quite the same as phosphorus-sensitive plants but considerable similarities occur, so much so that low-fertility acid soils are specifically designed to meet their needs.

Most fertilisers emphasise phosphorus in readily available soluble form. Also, manures, particularly poultry manure, have high levels of soluble phosphorus and may cause toxicity in P-sensitive plants. Products such as 'blood and bone' or compound fertilisers based on rock phosphates are safer because the phosphorus is not soluble. Low P 'native plant food' fertilisers can be purchased, which have low solubility forms of P.

pH affects phosphorus availability as well as levels of iron in soils, with acid soils being less prone to acute P toxicity.

Species with proteoid roots (i.e. the Proteaceae) are particularly sensitive to soluble phosphorus in the soil.

3.3.2 High-salt-tolerant plants

Plants vary greatly in the amount of salt they can tolerate, and some have evolved to grow in particularly high salt conditions. There is readily available information on the salt tolerance of plants, and it is often part of a plant profile description in various texts. Plants naturally occurring in coastal areas generally are more salt tolerant.

As well, species naturally occurring on swampy, poorly drained soils and besides riverbanks can usually tolerate inundation by water. It is important to analyse the site and its context to find clues to the conditions so that they can be considered in the landscape design approach.

Table 3.3 lists some commonly used landscape plants and their tolerances. This list is adapted from Handreck and Black (2010).

Table 3.3.   A general selection of plants adapted to specific soil attributes

Soil attribute	Plant
Acidic soils (pH: 5.2–6.5) Plants preferring acid soils (iron-inefficient plants)	Ericaceae family: *Azalea, Rhododendron, Pieris japonica* Theaceae family: *Camellia* Rutaceae family: *Boronia* (citrus pH 6.0–7.0), *Gardenia, Murraya* Begoniaceae family: *Begonia* Hydrangeaceae family: *Hydrangea macrophylla* and *H. serrata* will go blue due to available Al ions at pH below 7. Edibles: blackberry, blueberry, broccoli, corn, cucumber, beans, onions Trees deciduous: *Fagus* spp. (beech), *Acer* spp. (maples), *Salix* spp. (willow), *Quercus palustris* (pin oak), *Betula* spp. (birch), *Arbutus unedo* (strawberry tree), most *Magnolia* spp., most *Pinus* spp., *Liquidambar styraciflua, Cornus* spp. (dogwood) Trees evergreen – Australian: *Acacia melanoxylon* (blackwood), *Backhousia anisata* (aniseed myrtle), *Callistemon* spp. (bottlebrush), *Telopea speciocissima* (waratah)
Alkaline soils Plants tolerant of alkaline soils (pH: 7.0–8.7) Iron-efficient plants These plants may also display symptoms of iron (Fe) and manganese (Mn) deficiency in alkaline soils.	Perennial plants: lavender, iris, *Sedum, Nepeta, Aquilegia, Hemerocallis, Delphinium, Kniphofia, Salivia, Chrysanthemum, Miscanthus, Clematis* Annuals: *Calendula, Alyssum* Shrubs: *Thuja, Cotoneaster, Euonymus, Chamaeocyparis, Hibiscus, Spirea, Vibernum, Taxus* Hydrangeaceae family: *Hydrangea macrophylla* and *H. serrata* will go pink at pH above 7 Edibles: tomato, spinach, cauliflower, sweet pea, kale, asparagus, brussel sprouts, garlic, pumpkin Trees deciduous: Select maple species only: *Acer miyabei, A. ginnala* (Amur), *A. campestre, A. platanoides* (Norway), *A. griseum Aesculus carnea* (red horse chestnut) *Carpinus betulus* (hornbeam) *Cercis canadinsis* (redbud) *Cladrastis kentukea* (yellowood) *Cratageus crus-galli* (hawthorn) *Ginkgo biloba* (ginkgo) *Gleditsia tricanthos 'inermis'* (honey locust) *Koeluteria paniculata* (golden rain tree) *Malus* sp. (crabapple) *Parrotia persica* (parrotia) *Pyrus calleryana* (callery pear) *Quercus imbricaria, Q. macrocarpa, Q. robur, Q. muelenberghii* (oak trees) *Tilia cordata* and *T. tomentosa* (linden) *Ulmus parvivolia* (lacebark elm) *Zelkova serrata* (Japanese zelkova)

Table 3.3. (continued)

Soil attribute	Plant
	Trees evergreen: *Cedrus atlantica*, *C. deodara*, *C. libani* (Atlas, Deodar and Lebanon cedar) *Picea abies* and *P. omorika* (Norway and Serbian spruce) *Pinus bungeana* and *P. flexilis* (lacebark and limber pine) Trees evergreen (Australian): *Cupaniopsis anarcardioides* (tuckeroo) *Casuarina glauca* (she-oak) *Eucalyptus deglupta* and *E. burdettiana* *Acacias* (wattles) including *A. acinaceae*, *A. longifolia*. *A. pendula*. *A. pycnantha* Some *Melaleuca* species (*M. linarifolia*, *M.lanceolata*, *M. huegelii*, *M. pentagona*) Many other Australian plants: most hakeas, grevilleas, and banksias in lightly alkaline soils
Plants sensitive to phosphorus require low-phosphorus soils. (plants sensitive to P can sometimes also show Fe deficiency). Soils with high organic matter (10–15%) are often too high in P for P-sensitive species.	Proteaceae family: most *Banksia* species and syn. *Dryandra*, all *Grevillea*, all *Hakea*, and *Telopea speciosissima* (waratah) Proteaceae family (from southern Africa): *Protea* and *Leucodendron* Fabaceae family: *Bossiaea*, *Daviesia*, *Hardenbergia*, *Kennedia*, *Chorozema*, *Gastrolobium*, *Eutaxia*, *Jacksonia*, *Pultenaea*, and very many *Acacia* that have evolved on coastal sandstones and low-P sands (such as in south-west Western Australia) and coastal heathlands. Goodeniaceae family: *Lechenaultia* Some Myrtaceae family: *Baeckea*, *Beaufortia*, *Hypocalymma* Cunnoniaceae family: *Bauera* Some Rutaceae family: *Boronia* Many other Australian native plants, but not all, are very efficient at extracting phosphorus from soils and can uptake too much, resulting in shortened lifespans or toxicity
High salt content soils (plants known to be very tolerant to salinity in soils) EC saturated extract no higher than 13 dS/m).	*Acacia cyanophylla*, *A. cyclops*, *A. longifolia*, *A. pulchella* *Araucaria heterophylla* *Atriplex* spp. Some *Banksia* *Callistemon citriunus* *Carissa grandiflora* *Carpobrotus chilensis* and *C. edulis* *Casuariittateyla* and other *Casuarina* sp.
Note: soil leaching can reduce salts in soils	*Coprosma repens* *Cordyline indivisa* *Correa alba* *Delasperma* spp. *Eucalyptus camaldulensis*, *E. sargentii*, *E. ittateteata* (very tolerant 13 dS/m) *Eucalyptus botryoides*, *E. cooabah*, *E. occidentalis*, *E. pileate*, *E. robusta*, *E. sideroxylon* (tolerant 8 dS/m) *Ficus macrocarpa* *Hibbertia scandens* *Leptospermum laevigatum* *Leucophyllum frutescens* *Melaleuca armilaris*, *M. diosmifolia* and *M. nesophila* *Pelargoinittateale* *Phoenix dactylifera* *Rhagodia* spp. *Scaevola calendulaceae* *Westringia fruiticosa*

Soil attribute	Plant
Low salt content in soils (plants sensitive to salinity in soils) EC saturated extract no higher than 1.8 dS/m). (Refer to Section 3.3.2.)	*Acanthus mollis* *Camellia* *Cedrus atlantica* *Cotoneaster* *Feijoa sellowiana* Ferns *Gardenia* spp. *Ilex cornuta* *Phormium tenax* *Photinia* spp. *Podocarpus macrophyllus* *Rhododendron* spp. *Rosa* spp. *Salix purpurea* *Tilia cordata* *Trachelospermum jasminoides* *Viola hederaceae*, African violet, violet Edibles sensitive to salinity in soils: blackberry, strawberry, carrot, beans, *Persea americana* (avocado), also apple, stone fruit, capsicum, radish, lettuce

Note: the above lists are not exhaustive and intended as an informative guide only. Commonly specified plants have been listed.

3.4 SOIL CONTAMINATION

Soils contaminated with both organic and inorganic materials are sometimes present in urban areas and some rural environments (e.g. old stock-drenching yards or former mined sites or former industrial sites being re-zoned into residential developments).

It is not the intention of this publication to describe or specify what must be done to identify such contamination, because this is controlled by various local, state and federal environmental protection legislation.

Environmental rehabilitation and remediation is a highly specialised and regulated profession. Where any questions of site contaminations arise, it is recommended that an environmental contamination specialist is engaged to conduct the statutory investigations. Where levels of contamination above the published guideline levels are found, remediation of the site during development will take a high priority in the planning consent process and its feasibility.

The relevant state and federal authorities have lists of known previous site use categories that mandate statutory site investigations during the planning process. An example would be where the site used to be an industrial facility, smelter, waste tip or farming land. These will be published on the relevant authority's website. Even if this is not triggered, during the site investigation process the designer may come across anomalies that would trigger environmental investigation.

Landscape architects and designers should be aware of the common site observations that may prompt some additional contamination investigation. Observations of potential soil contamination include:

- the presence of oily or tarry matter or if the soil is bare or very poor plant growth
- the presence of intense garbic layers associated with past industrial or land-filling activities
- the presence of any unusual smelling or coloured matter
- solvent, chemical, or petrol smells

Table 3.4.   Common soil contaminants

Metal and non-metal contaminants	Organic contaminants
Arsenic	B-tex
Barium	Total recoverable hydrocarbons (TRH)
Beryllium	Polycyclic aromatic hydrocarbons (PAH)
Cadmium	Chlorinated pesticides (aldrin, dieldrin, heptachlor, chlordane, DDT, DDE, DDD)
Chromium	Cyanides
Copper	
Lead	
Mercury	
Nickel	
Zinc	

- any chlorosis, necrosis (leaf burn) or deformity of vegetation or unknown death or ill-health of plants
- the presence of calcification or salt stains
- chips and fragments of bonded asbestos cement
- any observations of high-level site disturbance.

Where such observations are made, the landscape architect or designer has a duty to inform the client that a contaminated site assessment under the relevant state environmental legislation may be required to assess the implications for human and environmental health.

> Knowledge of the type of soil contamination will in turn assist in determining the Soil Approach Method, landscape design and soil specifications.

The most common substances analysed for during such investigations are shown in Table 3.4.

The list in Table 3.4 is not exhaustive and if an environmental investigation is required the environmental specialist will advise on how to investigate further. Refer to your local, state or federal environmental protection requirements to see if statutory investigation is triggered.

3.4.1 Contaminants and vegetation

Generally, the only contaminants likely to impact vegetation are solvents, high oil and grease levels (hydrocarbons) and metals with some physiological role in plant biochemistry, such as manganese, zinc and copper.

Zinc and copper are the most common metal contaminants seen in urban sites at levels likely to affect plants. Other metals that interfere with plant biochemistry are nickel and, at very high levels, cadmium and lead.

That said, metal toxicities in plants are rarely seen. This is for two reasons:

1. Plants are generally quite insensitive to heavy metals. Most guideline levels are usually much lower than those that would affect plants. If you obtain a soil

contamination report that states there is contamination exceeding these guidelines, it does not necessarily, or even usually mean, it will detrimentally affect your plants.
2. Natural selection has occurred and there are no plants left growing on the site that are affected by the levels seen. Some grasses for example, are highly tolerant of metals and give the site the appearance of being well vegetated.

Lead (Pb), for example, is not highly phytotoxic and will generally only affect plants at very high levels as it is very insoluble and has no physiological role in plant metabolism.

Metals that do affect plants usually have a physiological role and include zinc (Zn) and coper (Cu). Elevated Zn and Cu are not uncommon to come across on urban project sites but we foresee the number of highly contaminated sites will diminish over time as they are remediated through the development process.

With the increase in in-fill residential developments due to increased urban pressures and densities we do see lower level contamination, particularly with Zn, Cu and Pb on older suburban sites as becoming increasingly common for the landscape architect to deal with.

Short-chain hydrocarbons (petrol and diesel) affect plant roots. Higher chain hydrocarbons (oils and greases) need to be at very high concentrations before a detrimental effect on plants is observed. This is in part because by their very nature they are water repellent. Highly volatile petrol and benzenes are toxic to all living things.

The essential point to remember is that the triggering of a human health or environmental guideline value does not necessarily mean the site is not suitable for plant growth. Additional investigations and advice may be required.

3.5 SOIL REMEDIATION

An existing site soil may be less than ideal for landscape development due to the presence of contaminants. One of the higher priority landscape soil approach methods (Section 4.7), and, incidentally, a high priority in all environmental guidelines, is to work with the existing site soil to avoid landfill and design the landscape with relevance to the soil constraints.

In site remediation, consideration of an alternative approach to soil removal to landfill should also be explored for project suitability, such as:

- chemical or physical remediation treatments (e.g. leaching of saline soil, treating with wetting agents and integrating organic matter)
- capping of contaminants at depth using clay and visible 'marker' layers to warn of contamination is very common. We encourage landscape architects to consider alternative approaches before accepting capping is the only option.
- phytoremediation using composts, fertilisers planted with resistant species to improve the rhizosphere, allowing plant roots to access the soil more deeply and restore a healthy biological condition to aid degradation and removal of organic contaminants
- staged management processes (such as phytoremediation, collecting concentrated bioaccumulated plant material for moving off site in a staged process with possibly another stage for soil improvement such as using green manure). Again, all of these considerations are completely project dependent.
- growing of salt-tolerant cover crops on salinated soils together with irrigation and leaching to desalinate surface soil before planting the permanent landscape plantings.

The use of on-site phytoremediation is of increasing interest in lightly to moderately contaminated sites. This approach usually takes time and planning, and in heavily contaminated sites, statutory obligations to remediate will often take precedence. However, it is worth encouraging the on-site remediation approach, noting some key points:

1. Certain plants may still grow well enough despite the contamination. Refer to Tables 3.5, 3.6 and 3.7.
2. Contaminated sites are sometimes poorly vegetated, not because of the contamination but due to compaction, poor soil chemistry, deficient nutrients, extreme surface temperatures and high run-off coefficients leading to a soil moisture deficit and hence a low soil biological presence.
3. Managing surface water flows to stop run-off and erosion by improving water infiltration and storage (water harvesting) will assist plant growth.
4. Aquatic plants enhance biofiltration filter run-off water, decompose contaminants like hydrocarbons and improve waterway health.
5. The very existence of plant communities, with their rhizosphere and moderation of climate at the surface of the soil, has a beneficial effect on the soil and soil microbial health.
6. Optimising plant growth at the soil surface will lead to faster metabolic and physical processes to degrade organic contaminants.
7. By the simple process of optimising soil physical and chemical conditions we can greatly increase the rate of natural remediation, particularly of organic contaminants.
8. Some plants are known to hyperaccumulate heavy metals in contaminated soil, but the process is slow and requires harvesting and removal of the plant material at regular intervals.

Generally called 'phytoremediation', the process of optimising plant growth will lead to faster soil remediation. These methods can often be of initial lower cost in comparison to conventional approaches to site remediation but yield less immediate results being much slower to establish, and slower to bring sites within regulatory upper limits for contaminants.

Where development has time limits, the conventional approaches of containment or removal is likely to be preferred by the client and the environmental consultants.

Table 3.5.   Key plant species typically used for phytofiltration in wetland, raingarden and water-edge settings

Plant (sedge, reed, rush, rhizome)	Type
Baloskion tetraphyllym, feather top (tassel cord rush)	Phytofiltration and stabilisation
Baloskion syn. *Restio tetraphyllus* (tassel cord rush)	
Carex appressa (common sedge)	
Dichelachne micrantha (short hair plume grass)	
Ficinia nodosa syn. *Isolepis* (knobby club rush)	
Juncus usitatus (common rush)	
Lepironia articulata twizzler (twisted reed)	
Lomandra fluviatal 'Shara' (mat rush)	
Pterittateata (brake fern)	

Waterways and biofiltration beds are often used to clean up or improve water quality. Key to this is the use of reeds and sedges. Plants commonly used to encourage 'phytofiltration' are listed in Table 3.5.

These reed, rush, sedge and rhizomatous plants, and one fern species are mostly native or cultivars to Australia. Given that these plants are typically used in constructed wetlands, raingardens, detention basins, and at the edges of waterways, it is advised to check local suitability and availability to protect local native ecosystems.

Consider also trees at the edges of waterways to assist in phytofiltration. Often suitable trees will be deeper rooted (up to 0.8 m depending on the soil type) than the sedge and rush species (typically only in the top 0.3–0.4 m). The combination of both tree and sedge plants is likely to be of more efficient benefit.

Some commonly used tree species in phytofiltration are listed in Table 3.6. Like the plants in Table 3.5, check the suitability of these tree species in relation to the project's local surrounding native context and ecosystem, and of course in relation to the soil fertility results.

This is not an exhaustive list and many tree species will tolerate edge of waterway conditions and provide benefits including embankment stabilisation; uptake of groundwater, soil toxins, salts and nutrients; encouragement of soil microbial activity; and aeration.

3.5.1 Bioavailability of metals

The bioavailability of contaminants to plants should be considered for several reasons such as including plants for consumption or urban farming, as well as determining if a plant species is likely to even be affected by a contaminant that may actually be unavailable for plant uptake.

Bioavailability typically diminishes with time. This is also dependent on soil properties, the most important being soil pH, clay content, organic matter and types of minerals present.

Table 3.6.   Some examples of tree species typically used for phytofiltration along water-edge settings, in wetlands and raingardens

Tree	Type and notes
Populus deltoides (poplar, cottonwood), *Populus tremula* (quaking aspen).	Phytofiltration. Fast ground water uptake. Assists in accumulating chlorinated solvents (trichloroethylene and tetrachloride) and petroleum hydrocarbons (benzene, toluene and o-xylene).
Salix viminalis and Alba (Willow, white willow tree)	Phytoaccumulation, phytostabilisation or phytoimmobilisation. Assists in copper (Cu) uptake, heavy metals and diesel, cadmium (Cd), nickel (Ni), and lead (Pb). Suitable for growing beside wastewater bodies.
Paulownia spp. (Princess tree)	Phytoaccumulation. Fast-growing tree and effective in phytoremediation (set this tree back from the water's edge). Drought tolerant once established and suitable for ephemeral raingardens.
Melalecua quinquinervia (Paperbark)	Phytofiltration. Effective at growing along drainage channels, swales and raingardens.
Eucalyptus camaldulensis (River Red gum)	Phytofiltration. Drought tolerant once established.
Casuarina cunninghamiana, C. glauca, Allocasuarina torulosa (she-oak and river she-oak)	Phytofiltration and phytoremediation. Tolerant of some soil salinity.
Avicennia marina (Grey mangrove)	Phytofiltration. For coastal, intertidal waters' edges, consider the site's locally growing mangrove tree species.

Other factors determining the actual concentration of absorbed contaminants include the rate of plant growth and the climatic conditions (particularly soil moisture content). These can be assessed with the advice of a soil scientist or soil agronomist; however, predicting metal bioavailability can be highly error prone, and a bioassay or the measurement of metal contaminants in plant tissue actually growing on the site is more reliable.

Various plants accumulate toxins (such as metals) in only select parts (e.g. leaves, stems or roots). For example, lettuce, spinach and silverbeet tend to accumulate metals in their leaves, while potatoes accumulate metals in their tubers. Fruit usually shows much lower levels of contaminants than leaf tissue. Refer to Tables 3.7 and 3.8.

Additionally, it is important to mention that tilling and working a landscape and actively growing plants, and adding organic matter on a site together with added watering of that landscape and in turn improving soil microbial activity, can vastly improve the growth of plants and, if contaminants are bioavailable, the typically slow process can be sped up, but it will almost always be a very slow process. There are other reasons to implement remediation including retarding spread or movement of contamination in soil (both outwards and downwards), which could be important regarding proximity to waterways, catchments or other nearby 'susceptible' land uses.

Table 3.7. Examples of plants grown for consumption (edible plants) which can take up toxins, but not always into the edible part

Plant	Description – edible part tends not to be affected by toxin
Tomatoes	Tend not to accumulate metals in their fruit.
Carrots and potatoes	When grown in soil with high concentrations of lead (Pb) and cadmium (Cd), they absorb these metals but amazingly tend not to let these metals pass through their epidermis (outer layer). Removing (peeling) the epidermis and moving and collecting this concentrated material (for contained storage that doesn't re-enter the soil zone) can start to remediate the site soil.
Rosemary	Leaves tend to be unaffected and not contain metals from soil.
Tagetes erecta (marigolds)	Tolerant of Pb, Cd and Zn and a hyperaccumulator of Cd and Zn but typically only in the roots. Flowers tend not to accumulate (noting their attractiveness to insect pollinators). Marigolds have good uses as companion plants.

Table 3.8. Examples of plants grown for consumption that tend to be affected by toxins and it is advisable to not consume if growing them in contaminated soils

Plant	Description – edible part tends to be affected by toxin
Amaranths (e.g. spinach, silverbeet)	Tend to accumulate metals in their leaves.
Brassicas (e.g. cabbage, broccoli, kale, mustard, cauliflower)	Tend to accumulate toxins in their leaves and stems.
Corn	Tends to absorb toxins (e.g. Pb, Ti (titanium) and Cu) which can impact the cob development.
Helianthus annuus (common sunflowers)	Tend to concentrate (bioaccumulate) toxins in their seeds. Famously known to remove radioactive uranium isotopes around nuclear plants such as in Ukraine. Also accumulates zinc (Zn) and lead (Pb).
Pumpkins	Tend to hold lead (Pb) commonly found in old lead-based house paints.
Lettuce	Tends to absorb Ni, Co, Fe within its leaves.
Rhubarb	Tends to be impacted by chromium and magnesium (plant stems do not develop typically and stem colouring can be impacted).

Urban farming on contaminated soils occurs widely in Europe and rarely leads to human health issues. Where high concentrations of soil contaminants are known, the avoidance of edible plants known to accumulate toxins in their edible parts (such as outlined in Tables 3.7 and 3.8) is a sensible precaution. Instead consider alternative urban farming methods in these instances, such as building up planters above ground, or separating planting zones from contaminated areas (such as in containers). Many laboratories will offer a soil-testing package for contaminants in urban gardens and advice can be sought on the management of any contamination.

The intention of this chapter is to emphasise that contamination on a site does not necessarily mean all the soil must be disposed of or capped. There are other solutions and they should be explored with the relevant experts.

Most attempts at phytoremediation of soil are still in the experimental stage and occur on sites where public access is restricted, but for those seeking further information on this area of growing interest, a reading list is attached at the end of the chapter.

Working with specialists in environmental rehabilitation and remediation is highly recommended where any questions of site contaminations arise. Statutory investigations regarding contamination may also apply to the site.

3.5.2 Bioremediation relevance for landscape architecture

Landscape architects and designers need to be wary of the misperception of phytoremediation cleaning up a site.

Heavy metals are non-biodegradable and can only be transferred from one chemical state to another.

This might mean the heavy metals in soils can be stored in plants if the heavy metals in soil become bioavailable. The other transformations can be volatilised, more water soluble (and potentially more mobile in soils), or less water soluble (locked up) such as with plants that hyperaccumulate heavy metals.

There are three main ways that specific plants 'deal' with bio available soil contaminants:

1. Plants can take up and store the toxin (bioaccumulation, hyperaccumulation, phytostabilisation).
2. Plants can transform contaminants in their plant tissues (phytodegradation, biodegradation).
3. Plants can metabolise contaminants in surrounding soil (rhizosphere biodegradation) with microbes and plant metabolisation, using sunlight energy and air (such as oxygen).

'Phyto' means relating to plant(s), 'rhizo' relating to root(s) and 'bio' (biological) relating to life and living beings.

Specifying plants that are tolerant of surviving in contaminated or susceptible environments is a suitable approach in some projects. Interplanting plant species to assist the growth of other plants (such as in companion planting practices) can also be a useful approach to landscape design as can deploying phytofiltration plants. Refer to Tables 3.5 and 3.6.

There are other benefits for implementing phytoremediation practices in landscapes including for land stabilisation, managing surface water flows, water filtration/biofiltration and watertable control.

Phytoremediation applications can also be considered to:

- support overall landscape function (through interplanting and companion planting)
- slow, retard or reduce the effects of contaminated sites or migration of contaminants (albeit slowly)
- lower maintenance inputs
- improve establishment of landscapes
- improve landscape longevity
- be a lighter approach to alternative (conventional) landscape interventions.

Phytoremediation applications can also, however, be less reliable; the benefits are harder to quantify and slower to establish than conventional methods such as strip and remove or cap over, but nonetheless landscape architects should still advocate for such applications in certain projects towards regenerative and/or sustainable landscape approaches.

3.5.3 Further reading – in situ phytoremediation topics

Boi J (2015) *5 Best Plants for Phytoremediation*. Land8 Landscape Architects Network, <https://land8.com/5-best-plants-for-phytoremediation/>, accessed 20 March 2024.

Dhir B (2013) *Phytoremediation: Role of Aquatic Plants in Environmental Clean-up*. Springer, New Delhi.

Health Guide (2022) *List of 10 Plants Used in Phytoremediation*, <https://healthguidenet.com/foods/plants-used-in-phytoremediation/>, accessed 20 March 2024.

Kafle A, Timilsina A, Gautam A, Kaushik A, Bhattarai A, Aryal N (2022) Phytoremediation: mechanisms, plant selection and enhancement by natural and synthetic agents. *Environmental Advances* **8**, 100203. doi:10.1016/j.envadv.2022.100203

Lu J, Yuan M, Hu L, Yao H (2022) Migration and transformation of multiple heavy metals in the soil-plant system of e-waste dismantling site. *Microorganisms* **10**, 725. doi:10.3390/microorganisms10040725

Madanan MT, Shah IK, Varghese GK, Kaushal RK (2021) Application of Aztec Marigold (*Tagetes erecta* L.) for phytoremediation of heavy metal polluted lateritic soil. *Environmental Chemistry and Ecotoxicology* **3**, 17–22. doi:10.1016/j.enceco.2020.10.007

McCutcheon S, Schnoor J (2003) *Phytoremediation, Transformation and Control of Contaminants*. John Wiley & Sons Inc, Hoboken, NJ.

3.6 SITE SOIL SURVEY

It is strongly recommended that the landscape architect or designer engage specialist soil experts for the soil component of site analysis to ensure that sufficient information is collected to inform the design process properly.

The soil profile layers may not be obvious to the unpractised eye and the services of a skilled pedologist are justified on larger projects. Geotechnical experts and engineers generally do not evaluate soils from a landscape and fertility perspective in the same way a landscape technologist does. Relying on geotechnical engineering reports to provide soil information for design purposes seldom provides a satisfactory outcome, because their focus is on engineering properties.

A survey of a site's soil resource should be conducted to identify, as a minimum:

- the depth of each soil horizon
- the morphology (texture, structure and colour) of at least the A and B horizons
- the presence of any inclusions (ironstone, manganese pellets, lime concretions)
- the soil type or classification of the soils present
- any areas of disturbed, filled or altered conditions that render the soil unusable or raise special requirements
- the depth of the topsoil and any variation in depth for stripping purposes.

Investigation of soils involves digging to around 600–1200 mm depth and exposing the surface horizons of soil. As a minimum, the topsoil (A horizon) and subsoil (B horizon) should be exposed.

If salinity is suspected, then at least some bore holes should be made to the C horizon (> 1200 mm).

It is usual to use a hand soil auger of 50–75 mm diameter, but, on large sites or where depths greater than 900 mm are to be investigated, the use of a backhoe greatly aids the process. Backhoe pits give greater opportunity for examining intact soils and taking photographic records.

3.6.1 Site analysis – soils summary

Table 3.9 outlines general characteristics and observations required in site analysis phase. This table may be used as a checklist for experienced industry professionals or as a 'scope of works' list for urban landscape soil scientists for field analysis.

For a site soil checklist and further details see **Appendix B**, Section B1.

Refer to **Appendix A** for further information on soil survey, sampling procedures, and soil-testing methods.

Refer to **Appendix B** providing site soil analysis checklist, on site compaction test, and example soil analysis soil chemistry reports.

Table 3.9.   Site analysis observations

1. General site	Slope aspect Slope position
2. Vegetation	Species and condition of vegetation Degree of alteration from natural condition Presence and condition of any weeds Degree of stress and/or disease
3. Surface conditions	Grass and forbs Intact litter layer Crusted or compacted surface
4. Topsoil	Depth Colour, texture, structure Moisture condition Presence of inclusions (anthropic objects, Fe/Mn nodules, stone) Degree of compaction Surface cracking and crusting Presence of any pallid layer (A2 horizon) on top of the subsoil
5. Subsoil	Depth of the boundary Depth of subsoil Colour, texture, structure Colour and texture changes to the deep subsoil Depth to parent material (not always possible)

3.7 SAMPLING AND ANALYSIS

We encourage landscape architects and designers to inform their designs by engaging with soil specialists who will provide key information that is essential to shape plant selection, layout, landform and general landscape design.

While investigating the site, it is usual to take soil samples for laboratory testing. Take samples of each layer or horizon at each bore hole or inspection pit and record the sampling location and depth of each sample. For higher level or larger scale projects, the landscape architect may need to engage with a soil scientist early in the soil design process.

The number of samples that will be tested and the type of tests that need to be performed will depend on the budget, complexity and cost of the landscape works, the uniformity of the site soils, site size, topography and many other factors. The following points illustrate how sample numbers can be contained by creating bulk or composite samples together. Composite soil samples can be performed where:

- topsoils show similar properties and that come from the same topographic position (e.g. a composite sample created out of all the footslope samples from one area)
- subsoils are of the same colour and texture (e.g. all red clay subsoils from the same depth and topographic position could be mixed together to form a composite soil sample).

The number and location of bore holes or test pits required to represent the majority of the soil properties on a site will vary with soil type and uniformity. Note that the location of bore holes to obtain representative samples is important because soils vary mostly with slope position. The following general statements apply:

- On flat sites, spread locations approximately evenly.
- On sloping sites, make a catena of sample locations from top of hill or crest to the bottom of the hill or floodplain.
- On sites of variable slope, treat areas of the same slope as likely to have similar soil (e.g. a floodplain is likely to be one soil type and a footslope another).

To gain more information on topsoil depths, additional 'confirmatory' holes can be made in greater numbers. Note: always fill test holes back in the correct order so that the soil from the lowest depth goes back in the same place as it was and the topsoil ends up in the topsoil layer.

Appendix A and **Appendix B** provide additional tables on soil sampling, compositing and examples of 'Soil test request form' (Fig. B3) and example 'Soil test results'. The development of a complex sampling plan for assessing soil resources requires the skills of a professional soil scientist.

Individual laboratories will vary in the range and naming of test procedures so it is best to contact them and establish what you need. Many laboratories will not interpret results, so either take them to an experienced agronomist or soil scientist or find a laboratory that will interpret them and provide recommendations based on the intended landscape design application.

3.7.1 Laboratory analysis

A horizon (topsoil) and B horizon (subsoil) should be analysed for the properties shown in Table 3.10 as a minimum. It is also advisable to sample and test the C horizon (deep subgrade) where cut and fill may bring this to the surface for potential use as a post-construction subsoil or subgrade.

Table 3.10.   Minimum analysis of properties

A horizon or topsoil to be analysed for (at a minimum)	Subsoil and subgrade to be analysed for (at minimum)*
pH	pH
Salinity (electrical conductivity)	Salinity (electrical conductivity)
Cation exchange properties and exchangeable cations	Cation exchange properties and exchangeable cations
Major and minor nutrients	Aggregate stability (Emerson aggregate class)
Organic matter (%)	
Particle size analysis**	

*Where there is any suspicion of salinity, it is strongly advised that a deep subsoil sample, to around 800 mm depth, be taken and also analysed for subsoil properties.
**Particle size analysis is only really necessary where the subsequent specifications call for a particle size distribution test (e.g. for sports field or recreational turf specifications). In most cases field texture and structure assessment are adequate for prediction of properties such as drainage.

3.8 REPORTING ON SOIL CONDITIONS

3.8.1 Findings and recommendations

Test results and the results of the soil survey may need to be written up into a report describing the findings, implications for landscape development and recommendations to the designers and specifiers. Refer to **Appendix A1** Taking *in situ* soil samples, and **Appendix B1**, Table B1 Site soil analysis checklist.

Such written reports may be within the skills of an experienced landscape architect or designer, or landscape contractor for smaller projects, but larger projects will normally require the services of a competent, trained soil scientist, agronomist or horticulturist experienced in soil science for landscape development or land rehabilitation.

The soil report should include detailed sections in order to identify the main elements of the soil profile that are relevant to the intended landscape purpose. Preferably, the report should use regional and national soil classification systems, and should provide clear 'plain English' descriptions, including: Example soil analysis results (see **Appendix B**, Fig. B2), and Soil test request form (see **Appendix B**, Fig. B3).

- a description of the field condition of the surface materials soil (results of the field survey)
- an interpretation of test results
- a statement of fitness for purpose as topsoil, subsoil or subgrade
- recommendations for reuse, amelioration, improvement or burial as subgrade.

The landscape architect, designer or landscape contractor or client can then use the analysis information in two ways:

1. Examine its impact on design considerations, even altering or modifying the design to accommodate shortfalls or other problems
2. Develop tender specifications for soil stripping and reuse including methods for ameliorating or improving the soils for the intended use. Ensure engineering and/or landscape 'plans with preliminary specifications' either contain or refer to soil stripping, stockpile locations and the landscape soil recovery specification.

3.8.2 Soil results interpretation

The analytical results of the soil investigation should be interpreted by a competent experienced soil or agronomy professional. The best single source guides to the interpretation of test results in Australia are Handreck and Black (2010), Peverill *et al.* (1999) and Hazelton and Murphy (2016).

The interpretation of soil results should focus particularly on identifying the following aspects:

- the correction of unsuitable pH
- the presence of salinity levels that may limit plant growth/health
- the presence of cation exchange anomalies, such as sodic, calcic, magnesic or aluminium toxic conditions
- the nutrient status and its suitability for the intended landscape plantings
- the potential for toxicities due to phosphorus (in P-sensitive plants), manganese, aluminium and zinc in particular
- the physical properties, particularly the potential for low permeability, resistance to compaction and potential for waterlogging
- the degree of stoniness that may affect reuse
- the amelioration of recovered soil to bring them within the specifications required by the landscape aims.

3.9 THE 'SOIL APPROACH' PROCESS

The 'soil approach' process is a term coined to describe a series of decisions or choices that are made as to:

- what the *source of soil* for the project will be (recovered soil, imported soil or a combination). Refer also to Chapter 4 and Table 4.1.
- what type of *soil profiles* are needed (profile form, refer to Section 4.2.2 and Table 4.4)

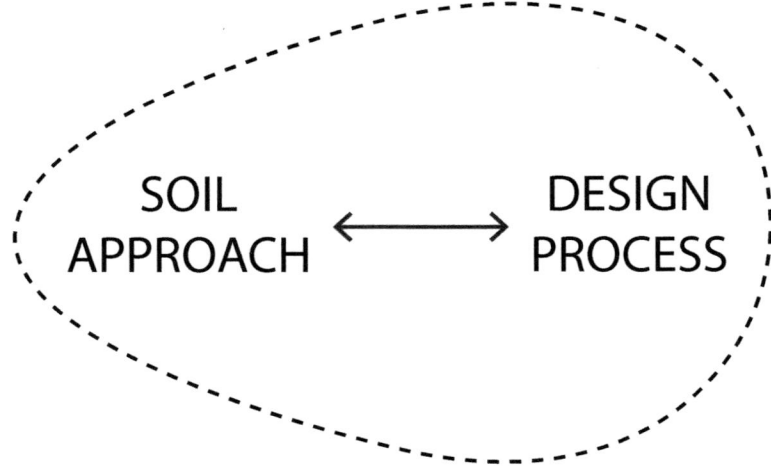

Fig. 3.1.   The soil approach interacts with the design and installation process, and both influence each other.

- what *depth* they will need to be for the chosen planting (refer to **Appendix C,** Table C2)
- what *performance specifications* will be needed for their construction (Chapter 6)
- what *quality assurance and control* processes will be appropriate to the scale of the project (Chapter 6, Part IV, G Specifications)

All this information is required for tenderers to calculate quantities and costs.

The soil approach interacts with the design process, and both influence each other in a 'back and forth' manner. Refer to Fig. 3.1.

3.10 BRINGING THE SOIL SELECTION PROCESS AND DESIGN TOGETHER

The flow chart in Fig. 3.2 illustrates a typical project process shown alongside the soil selection process. Landscape project processes can differ, but an attempt has been made to illustrate the major phases of the project design in relation to the soil design processes needed to properly realise those landscape phases.

3.11 IDENTIFICATION OF 'FIT FOR PURPOSE'

A clear statement of fitness for purpose of the soil should be documented. This is outlined in example **Specification G3**. The report should state whether the current site materials can be used in the landscape works for one or more of the following three uses:

1. Site material can be ameliorated for use as topsoil.
2. Site material can be ameliorated for use as subsoil only.
3. Site material can be used as subgrade only.

THE PROJECT DESIGN PROCESS ALONGSIDE THE SOIL DESIGN PROCESS

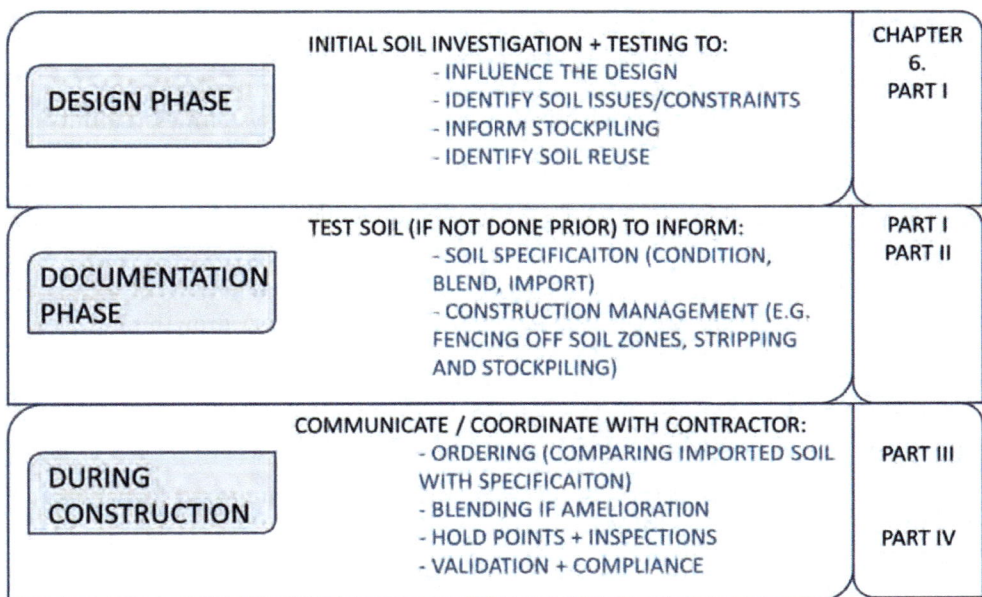

Fig. 3.2.   The major phases of the project design and installation alongside the soil design processes.

The interpretation of results shall be used to identify whether the site soils are suitable for the intended purpose.

The main issues that could result in a conclusion of unsuitability of site soils are:

*   the landscape design may be unsuited to the soil type (consider an appropriate alternative landscape concept for the site)
*   the stone content is too high
*   the soil has an unsuitable texture and structure to ensure adequate drainage (e.g. heavy clay)
*   the soil has an unsuitable texture for areas receiving heavy traffic (design solutions are therefore needed)
*   salinity and/or sodicity is too severe to be rectified
*   alkalinity is too high to be rectified by any practical means and/or the plant list cannot be sufficiently adapted to cope
*   the soil is contaminated with metals (particularly zinc and copper)
*   phosphorus levels are too high to be rectified for the growth of P-sensitive components of the landscape (design process to consider use of P-tolerant plant species. Refer to Table 3.3.)
*   the likely subsoil hazards include salinity, sodicity or extreme pH that may pose issues for its reuse as subsoil.

Refer also to **Specifications A1** and **G3**.

3.12 AMELIORATION AND IMPROVEMENT

Where site soil is determined as suitable, the current subgrade could be recovered and reused as part of a functional soil horizon (A or B horizon – topsoil or subsoil). A stripping and recovery plan should be produced documenting, as a minimum:

*   the depth or range of depths at which stripping of topsoil should cease
*   whether there is any need to strip and recover subsoil for reuse (it may be necessary to recover some subsoil where significant cut and fill will result in subgrade surfaces that are unsuitable for use as subsoil due to stoniness, salinity or other extreme chemical or physical hostilities)
*   the approximate volume of topsoil that is recoverable
*   a means of stockpiling that does not result in sediment loss from the site (such as over-sowing).

The interpretation of results should be used to identify any ameliorants or fertilisers required to improve landscape performance and obtain satisfactory growth rates. Table 3.11 provides a summary of the most common fertilisers and soil ameliorants used in landscaping. Further detail is provided in Handreck and Black (2010).

3.12.1 Why does soil fertility need improvement?

The question often arises as to why soil fertility needs to be improved for landscape purposes. This arises for several reasons:

*   The natural soil fertility is insufficient to support the landscape intentions. Sandstone-based soils, for example, will not support most landscape types except locally native, low-nutrient vegetation that is adapted to these

low-fertility soils. Most other 'garden' plants will starve without fertility enhancement.

- To maintain good turf coverage and recovery from wear in a heavily used urban setting requires the maintenance of good soil fertility. Starving turf of nutrients and moderate quantities of water will wear it out, since turf in active recreational use also requires good permeability and drainage to withstand compaction.
- Commercially available soils are often not topsoil, but are mined alluviums and need to be brought up to a topsoil fertility standard.
- The soil may have been degraded by the previous activities and lost its topsoil, been eroded in the past, or has been depleted of nutrient resources from previous vegetation and previous management practices on the site.

Table 3.11.   Ameliorants and fertilisers to improve landscape performance

Ameliorant or fertiliser	Function
Liming agents (lime, dolomite)	To adjust the pH of acidic soils upward, usually into the pH 5.5 to 6.5 range rather than neutral (pH 7)
Acidifying agents (sulphate of iron or agricultural sulphur)	To reduce the pH of alkaline soils, ideally to less than pH 7
Gypsum (calcium sulphate)	To ameliorate sodic and magnesic conditions and improve structure in calcium-deficient soils
Compound fertiliser (any particular combination of major (N, P, K), minor (Ca, Mg, S) and micro (Fe, Mn, Zn, Cu, B) nutrients)	To adjust fertility levels. Different N:P:K ratios can be chosen to emphasise one particular element (e.g. high N for promoting green foliage in plants)
Controlled-release fertilisers	To adjust fertility levels using a product that releases nutrients in a controlled manner over a stated period of time. Different N:P:K ratios can be chosen to emphasise one particular element
Straight N fertilisers	To emphasise only nitrogen, most commonly urea
Straight P fertilisers	To apply phosphorus only or mainly phosphorus (e.g. superphosphate, mono-ammonium phosphate)
Straight K fertilisers	To apply potassium only or mainly potassium (e.g. sulphate or potash and muriate of potash)
Organic matters (composts)	To improve soil organic matter, structure, texture and water-holding capacity. Composts vary in nutrient content
Manures	To improve soil organic matter, structure, texture and water-holding capacity and supply plant nutrients
Sands	Sand may be added to improve permeability and aeration, but it usually requires 60% sand or more to affect any appreciable change and more in heavily textured soils
Water-holding agents	Usually based on cross-linked polyacrylic polymers, these are most beneficial on very sandy soils of low water-holding capacity and are of doubtful value in soils heavier than loam texture
Wetting agents	Also known as surfactants, they are used to improve hydrophobic soils (soils that resist wetting when dry and also called water repellent)

3.12.2 Common ameliorants or enhancements

The most common ameliorations or enhancements used to improve soils for amenity horticulture are:

- the addition of lime or dolomite to ameliorate acidic pH
- the use of lime or gypsum to enhance exchangeable calcium and eliminate sodicity
- the addition of composts and/or manures to improve organic matter and nutrient levels
- the addition of single or 'straight' fertilisers to correct a deficiency of a particular element, for example the use of sulphate of potash to correct a potassium deficiency
- the addition of a multiple or 'compound' fertiliser to correct multiple element deficiencies including, where necessary, trace elements.

Treatments like these will often be required to meet the site soil specifications. Exactly what additions are required to meet them can only be determined by soil testing and interpretation by an experienced landscape soil technologist. Simply guessing and adding some of the common ameliorants or enhancements can quickly and easily cause greater deficiencies or more often toxicities, and lead to expensive rectification, plant replacement and poor landscape outcomes.

Also see Table 3.11 for further explanation of common ameliorants and fertilisers.

3.13 SPECIFICATION CASE STUDY 1: 'WOO-LA-RA', SYDNEY OLYMPIC PARK (SOP), NSW, AUSTRALIA

During construction of SOP (Fig. 3.3), a problem arose out of a complete lack of suitable topsoil for the remediation areas. The SOP site had large contaminated areas of soil due to previous industrial use. These areas involved the removal and reburial of millions of tonnes of old landfill, including the construction of the 'reburial mound' known as 'Woo-la-ra' (Fig. 3.4). The garbage was placed in a hill shape, then capped with highly imperme-able clay landfill capping. During a workshop with the design team from Hassell, Bruce Mackenzie and Peter Walker Partners, author Simon Leake explained how the natural soil landscape would have formed on such a hill and the edaphic relationships to vegetation.

The natural profile would have been a 'podsolic' type of soil with a sandy loam A horizon and a heavy clay B horizon. The topsoil would be only ~100–150 mm deep on the top of such a hill, gradually increasing to 350 mm on the footslope. The principal con-tractor said they were able to source plenty of crushed sandstone from excavations all around Sydney. Testing had shown that the clay capping could be ameliorated and, because it was deeper than the statutory requirement, there was no objection from the environmental managers to cultivating the surface 150 mm of clay. Environmental management did not want trees on a landfill, however, and could not be persuaded that trees would not root deeply (see Glossary: Root plate).

The author then suggested that a shallow 'facsimile' podsolic soil made from amelio-rated capping clay as the B horizon and screened crushed sandstone mixed with compost as the A horizon would provide a sustainable solution for the construction of a native grassland similar to the elevated plateau of open grassland of the former armaments depot nearby, which was at the time being demolished to make way for the Newington Olympic Village. At the bottom of the hill, the environmental team were content to allow trees where the topsoil would be deeper (at 500 mm). Thus, a naturalistic design was conceived, with open grassland 'rooms' providing some of the few open views of Sydney available on the site, and 'walls' of trees marching up the hills. The soil profile graduated

Above left: Woo La Ra between the Wanngal Woodland and Hill Road, nearing completion. Later to be covered in native Australian grasses.

Above centre: Haslams Creek with spot soil remediation in process. Note the bland condition of the engineered creek before the changes occurred.

Fig. 3.3. (Left) 'Woo-la-ra' between the Wanngal Woodland and Hill Road, nearing completion. Later to be covered in native Australian grasses. (Right) Haslams Creek with spot soil remediation in process. Note the bland condition of the engineered creek before the changes occurred. (Source: with thanks to Bruce Mackenzie Design.)

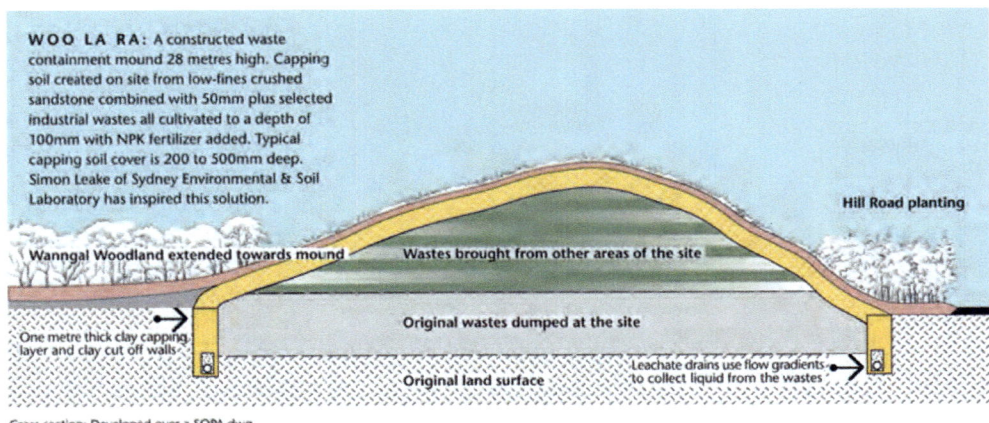

WOO LA RA: A constructed waste containment mound 28 metres high. Capping soil created on site from low-fines crushed sandstone combined with 50mm plus selected industrial wastes all cultivated to a depth of 100mm with NPK fertilizer added. Typical capping soil cover is 200 to 500mm deep. Simon Leake of Sydney Environmental & Soil Laboratory has inspired this solution.

Hill Road planting

Wanngal Woodland extended towards mound

Wastes brought from other areas of the site

One metre thick clay capping layer and clay cut off walls

Original wastes dumped at the site

Original land surface

Leachate drains use flow gradients to collect liquid from the wastes

Cross section: Developed over a SOPA dwg.

Fig. 3.4. The cross-section of the soil profile and waste containment at Sydney Olympic Parklands 'Woo-la-ra' site with 1 m thick capped clay layer using site resources. (Source: with thanks to Bruce Mackenzie Design.)

from 200 mm on the hilltops with grass to 500 mm on the footslopes, with trees providing a sustainable, low-cost solution using wastes that would otherwise go to landfill. The solution overcame all the physical constraints imposed by the site, as well as remaining within the landscape objective concept of 'rooms with walls' and meeting the high-priority environmental objectives.

No topsoil was imported for the project, all soil being made on site from excavated, crushed and screened sandstone waste.

3.14 SPECIFICATION CASE STUDY 2: BARANGAROO, SYDNEY CITY FORMER PORT, NSW

Background

The Barangaroo Headland Park development in Sydney involves the construction of 6 ha of parkland on a completely altered industrial port facility. A naturalistic sandstone headland is being re-created, involving a range of landscape units including a hilltop plateau, terraced slopes and harbour foreshore walks.

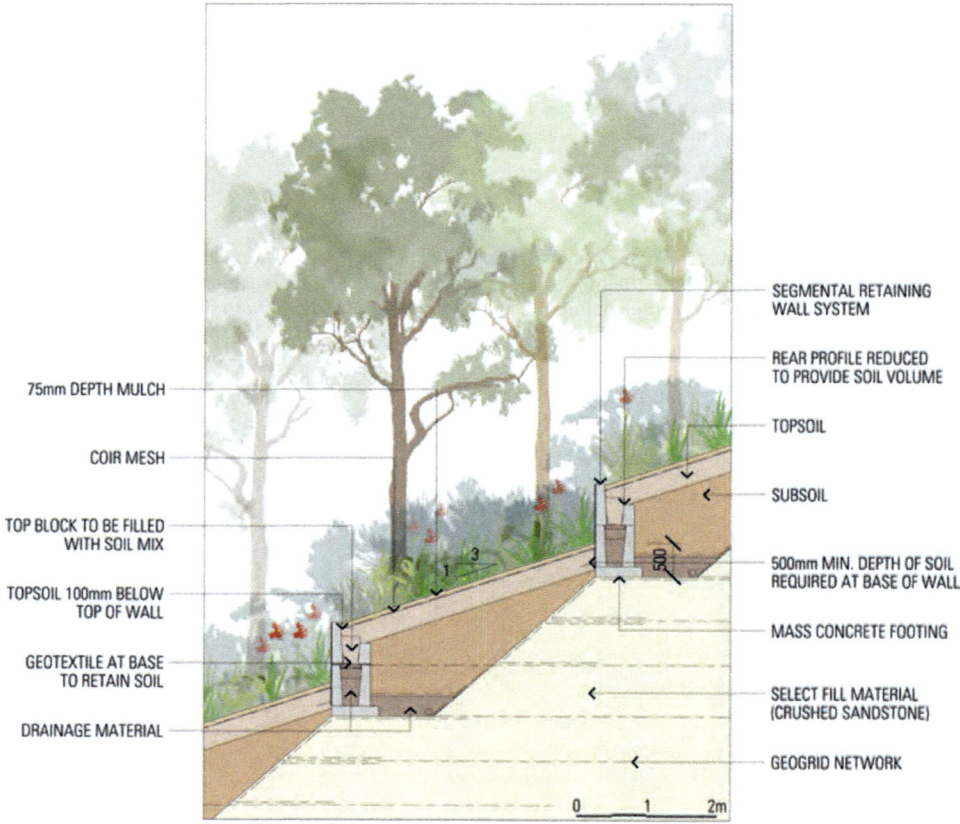

Fig. 3.5. The soil profile that was developed (over the carpark) for the Barangaroo Headland Park, 2013. (Source: Adam Robilliard and Adrian Pilton of Johnson Pilton and Walker and Barangaroo Development Authority.)

A range of soil profiles (e.g. Fig. 3.5) must be constructed suited to the establishment of amenity turf and trees in turf park areas, as well as areas devoted to the re-creation of representative Sydney Sandstone indigenous flora. There is limited research on Sydney Sandstone soils available largely because they have been neglected for agriculture. Some research is available from ecological studies on North Head and agricultural studies on the Kulnura Plateau north of Sydney.

No soil was available from the site, but a resource that was to be produced from the southern and central sections of the site was crushed sandstone from excavation. Using this as the basic raw material, research was conducted in four stages.

Stage 1. Sampling and analysis

Undisturbed soils in similar sandstone environments were sampled and analysed. This basically showed that phosphorus, calcium and manganese were the main limiting elements in the environment. Phosphorus was remarkably low, and analysis of plant tissue (foliage, twigs and wood) also showed levels much lower than that present in most plants (foliage around 0.06–0.12%, most plants being 0.2–0.35%). It was calculated that if all the P and Ca in the standing biomass were returned to the topsoil, it would show around 80 mg/kg P and 1000 mg/kg Ca. This is what would happen after a fire, so it was concluded that soils would have to start off at these kinds of levels for the vegetation to develop properly. Too much phosphorus can harm this low-nutrient vegetation type.

Stage 2. Physical properties

It was found that crushed sandstone has too much clay in it to drain properly and the natural soils are much sandier, at least in the surface topsoil horizon. For subsoil, it was found a mixture of 60% of 15 mm crushed sandstone and 40% of 3 mm washed sand gave acceptable permeability. For topsoils, the ratio varied from 30/70 for trafficked turf areas to 50/50 for native plant beds and the terraces not subject to traffic. All topsoils would need organic matter to be added, but not subsoils because organic matter is absent in subsoils of this type.

Stage 3. Chemical analysis

Analysis of sandstone, sand and locally available green waste compost was performed and used to calculate that around 10% compost by volume would be all that was required to mimic a low-fertility sandstone topsoil immediately following fire (the 'ashbed' effect).

Stage 4. Plant growth trials

Several trials were conducted using indigenous flora, including eucalypts, banksias and wattles, with additions of compost of 5%, 10%, 20% and 30% to the soil. These trials resulted in several basic conclusions:

- Addition of compost at 5% and 10% provided all the fertiliser needed for the lower status heath and open woodland landscape types. Any more than this resulted in excessive phosphorus levels and trace element deficiency.
- Addition of 20% compost would be sufficient for tall wet gully forest and trees in turf.
- Some additional nitrogen, but not other fertiliser, may be needed, at least for initial growth.
- pH adjustment using iron sulphate to bring pH down to around 5.5 was necessary due to the high pH of the compost (around pH 8).

Stage 5. Development of the specifications

In conjunction with Johnson Pilton Walker, technical performance-based and product specifications were drawn up. This occurred in three drafts, with some last-minute adjustments after the documents were published for tender. The main last-minute adjustment was to reduce the compost down to 5% for 'Type B soil' for the 'Ridge Top Woodland Heath & Scrub' landscape unit.

Excerpts of the tender-ready version of the specifications are reproduced above. The reader will note they are not exactly in the same form as that recommended in this text (which supersedes this case study) but contain the same components.

General background and 'Fit-for-purpose' statements

(Source: Reproduced with the permission of Adrian Pilton and Adam Robilliard, Johnson Pilton Walker and the Barangaroo Delivery Authority.)

General background

The Headland Park site contains no topsoil or subsoil and all materials will need to be manufactured from either recycled fill materials (e.g. crushed sandstone) produced on-site or imported quarry materials. These specifications describe the installation of a subsoil layer to all areas and then placement of one of three specifically designed topsoils to the depths required for the landscape installation. Each specification generally requires a drainage sand layer, a subsoil layer, a topsoil layer and either turf or a mulch layer.

Research by the Principal has demonstrated that the Sydney Sandstone flora components of this planting have specific nutrient requirements that are not met by conventional commercial soils. The Principal has found that cost effective alternatives can be used based on crushed sandstone and washed quartz sand derived from crushed sandstone together will green waste derived compost.

The specifications are presented in a form that summarises the compulsory (normative) set of chemical and physical performance specifications and then provides example formulations of components and fertiliser/ameliorant additives shown by the Principal to meet these performance specifications. The formulations are not compulsory and are given as informative advice; however, any tender desiring to produce alternative mixes must be able to demonstrate compliance with the compulsory performance sections.

Specifically excluded are calcareous recycled sand or sandy loam materials derived from building excavation in Sydney's eastern suburbs and Botany area. Much of the plant material in this landscape is of local indigenous origin and hence is known to be, or suspected of being, susceptible to phosphorus toxicity, iron deficiency in alkaline soils and/or is known to be adapted to low fertility acidic and highly drained soils. The Contractor must familiarise themselves with the nutritional and soil requirements of the vegetation type being planted.

The four soil types to be supplied are:

- Type A Headland and Foreshore Turf and Park Trees
- Type B Ridge top Woodland Heath and Scrub

- Type C Open Dry Forest, Tall Open Forest, Tall Moist Forest, Damp Gully Forest
- Type D Subsoil all areas.

These soil types are designed to suit the vegetation types and conditions.

Topsoil
Imported manufactured soil composition
The following tables provide a description of the four soil types and a recommended formulation that has been demonstrated to meet the specification. Alternative formulations will be considered provided compliance with the corresponding performance specifications of Tables below can be demonstrated.

Generally the fertiliser additives required to achieve the required levels of nutrients in topsoil will be:

- Green waste-derived compost: 5–20% depending on selection and chemical composition.
- Sulphate of iron: 0.3–1.5 kg/m^3 depending on the selection and soil properties.
- Slow-release nitrogen as IDBU or methylene urea: 0.5–1 kg/m^3 depending on selection.

Table 1.   Depths for Barangaroo

#	Item	Total soil depth including drainage layer (mm)	Topsoil depth (mm)	Subsoil depth inclusive of drainage sand layer (mm)	Drainage sand layer (mm)
1	Turf areas and selected garden beds	500	200	300	50–75
2	Trees	800	400	400	50–75
3	Fig trees	1500	400	1100	200
4	Terraces with shrubs and trees	1000–1800 (average 1400)	200–400 (average 300)	800–1400 (average 1100)	150

Technical specifications for Barangaroo Normative
General
Deliveries: Documentation to Australian Standard AS 4419, and Tables 2–4 (Barangaroo).

Additives: If using additives to raise topsoil to the required standard, ensure compliance with the relevant test criteria of Australian Standard AS 4419.

Nitrogen draw-down: If the NDI150 value is < 0.5 to Australian Standard AS 4419 Appendix E add a source of soluble nitrogen to bring the value above zero.

Compost: Provide well-rotted vegetative material or animal manure, free from harmful chemicals, grass and weed growth to the organic content by mass noted in the Selections (Table 5).

Performance specifications
Imported topsoil and subsoil
Particle size: Provide soil to the Particle size table for the vegetation types nominated in the Selections.

These physical properties will likely be met using the soil component formulations of Table 5 (below). However, the exact ratios of components and composts may vary with

Table 2. Topsoil physical properties ranges for Barangaroo

Sieve aperture (mm)	% passing by mass	
	Topsoil	Subsoil
15	100	100
5	90	90
3.35	80–90	80–90
2.00	70–80	70–80
1.00	50–70	50–70
0.50	35–50	35–50
0.25	20–35	20–35
0.15	15–20	15–20
0.106	12–25	15–20
0.053	10–20	15–20

Table 3. Subsoil chemical properties (Type D subsoil)

Determinant	Unit	Acceptable range
pH 1:5 in water	pH units	5.6–6.3
pH 1:5 in $CaCl_2$	pH units	5.2–6.2
Electrical conductivity 1:5 in water	dS/m	< 0.5
Total phosphorus	mg/kg	< 35
Organic matter	% w/w	< 0.5
Exchangeable sodium (Na)	% of CEC	< 10
Exchangeable potassium (K)	% of CEC	2–8
Exchangeable calcium (Ca)	% of CEC	45–65
Exchangeable magnesium (Mg)	% of CEC	25–35
Exchangeable aluminium (Al)	% of CEC	< 5
Mehlich 3 extractable iron (Fe)	mg/kg	100–300
Mehlich 3 extractable manganese (Mn)	mg/kg	25–50
Mehlich 3 extractable zinc (Zn)	mg/kg	5–20
Mehlich 3 extractable copper (Cu)	mg/kg	1–5
Mehlich 3 extractable boron (B)	mg/kg	0.5–5

Method references: Rayment and Lyons (2011), methods 4A1, 4B3, 3A1, 17B2, 15A1 and 18F1.

the source of soil product and type of compost. All formulations must be checked and validated as compliant with Table 1 before acceptance.

Nutrient levels: Provide soil to the Soil nutrient level (Tables 3 and 4).

Generally the fertiliser/ameliorant additions required to achieve these levels in subsoil will be:

- Gypsum: 1–2 kg/m^3
- Sulphate of iron: 1–1.5 kg/m^3

Suggested physical components and fertiliser/ameliorant additions required to achieve these levels in topsoil are given in Table 5. The exact ratios of components and

Table 4.   Topsoil chemical properties

Determinant	Unit	Acceptable
pH 1:5 in water Types B and C	pH units	5.6–6.3
pH 1:5 in CaCl$_2$ Types B and C	pH units	5.0–6.0
Electrical conductivity 1:5 in water	dS/m	< 0.5
pH 1:5 in water Type A	pH units	5.8–6.5
pH 1:5 in CaCl$_2$ Type A	pH units	5.2–6.2
Organic matter		
Type A	%	3–6
Type B	%	1–3
Type C	%	2–5
Total phosphorus		
Type A	mg/kg	90–150
Type B	mg/kg	50–80
Type C	mg/kg	60–90
Organic matter		
Type A and Type C	% w/w	2–5
Type B	% w/w	1–3
Cation exchange properties	% of CEC	< 10
Exchangeable Sodium (Na)		
Exchangeable Potassium (K)	% of CEC	2–8
Exchangeable Calcium (Ca)	% of CEC	45–65
Exchangeable Magnesium (Mg)	% of CEC	25–35
Exchangeable Aluminium (Al)	% of CEC	< 5
Mehlich 3 extractable Iron (Fe)	mg/kg	100–300
Mehlich 3 extractable Manganese (Mn)	mg/kg	10–25
Mehlich 3 extractable Zinc (Zn)	mg/kg	5–20
Mehlich 3 extractable Copper (Cu)	mg/kg	1–5
Mehlich 3 extractable Boron (B)	mg/kg	0.5–5
Nitrate-N (NO^3)	mg/kg	> 5

(Source: Rayment and Lyons (2011), methods 4A1, 4B3, 3A1, 17B2, 15A1 and 18F1.)

composts will vary with the source and type of soil, sand and compost components. All formulations must be checked and validated as compliant with Tables 2, 3 and 4 before acceptance.

Suggested formulations

Table 5.   Suggested formulations

Soil type crushed	15 mm sandstone % v/v	Washed 3 mm sand (% v/v)	Green waste-derived compost (% v/v)	Fertilisers/ ameliorants
Type A. Headland and	30	50	20	Fe2(SO4)3 1.5 kg/m^3 IBDU
Foreshore Turf and Park Trees				or methylene urea
				0.5 kg/m^3
Type B. Ridge Top Woodland	47.5	47.5	5	Fe2(SO4)3 1.5 kg/m^3 IBDU
Heath and Scrub				or methylene urea
				0.3 kg/m^3
Type C. Open Dry Forest, Tall	40	50	10	Fe2(SO4)3 2.0 kg/m^3 IBDU
Open Forest, Tall Moist Forest,				or methylene urea
Damp Gully Forest				0.5 kg/m^3
Type D. Subsoil, all areas	60	40	0	Fe2(SO4)3 0.5 kg/m^3

Sample submission and testing regime
Submissions
Requirement
All submissions to be made to the Principal's Representative within the timeframes specified in this Clause.

Soil tests for imported topsoil
Report: Submit a certificate noting the:

- suitability of each soil type for its specified use
- compliance with the normative selection specification
- suitability for establishment and ongoing viability of the site specified vegetation
- absence of any weed propagules or contaminants.

Submission time: 20 working days before installation of soil.

Site topsoil
Site topsoil salvaged from Munn Reserve may be reused on the Munn Reserve site if deemed suitable by Superintendent and if demonstrated as complying with the relevant

normative selection specification. It should be noted that additives and ameliorants may be required in order to demonstrate compliance.

Report: Submit a certificate noting the:

- suitability of the soil for its specified use
- suitability for establishment and ongoing viability of the site specified vegetation
- presence of any weed propagules or contaminants.

Submission time: 10 working days before installation of soil ON HOLD.

Samples

General: Submit representative samples of each material, packed to prevent contamination and labelled to indicate source and content for type testing bulk materials. Submit a 5 kg sample of each type specified for every 500 tonnes of imported soil. Submit bulk material samples, with required test results, at least 5 working days before bulk deliveries.

Submission time: 15 working days before installation of soil.

Suppliers

Statements: Submit statements from suppliers of soils and other materials, giving the following, where applicable:

particulars of the supplier's experience in the required type of work
production capacity for material of the required type, sizes and quantity
lead times for delivery of the material to the site.

Note: proprietary product specifications sheets will not be accepted as demonstrating compliance with the normative selection specifications.

Submission time: 20 working days before installation of soil.

Materials

Supplier's data: Submit supplier's data including the following:

- material source of supply for topsoil, filling, stone and filter fabrics. Compost: Submit a certificate of proof of compost compliance with Australian Standard AS 4454 Composts soil conditioners and mulches.

Submission time: 20 working days before installation of soil.

Execution

Program: Submit a work program in the form of a bar chart, for the installation of soil.

It is very important with such a sensitive vegetation type that no errors occur in the design, specification and enforcement of quality control. Such errors could be fatal to the landscape aims as many of the plants in the plant selection lists cannot grow, or are susceptible to constant insect attack when planted in excessively fertile soil, especially soil with excessive phosphorus levels. Phosphorus cannot be removed from soil and a legacy of high phosphorus would destroy the whole concept for this very prominent park development.

3.15 SPECIFICATION CASE STUDY 3: ONE CENTRAL PARK, BROADWAY, SYDNEY

The following extracts from the specification for One Central Park, Broadway in Sydney's CBD, Australia include performance specifications for a lightweight potting mix, which needs to provide longevity and be suitable for use in the planter boxes and rooftop garden beds on the high-rise buildings (as shown in Figs 3.6 and 3.7). One Central Park is a joint venture project by Frasers Property Australia and Sekisui House with design architect Ateliers Jean Nouvel of Paris and Patrick Blanc. As well, the project incorporates a Patrick Blanc-designed vertical wall for which the soil media had to be specified.

Source: Reproduced with the permission of Fraser's Broadway and Aspect Oculus Pty Ltd, landscape architects, Sydney. Note: this extract is not the entire soil specification for the project.

Performance specification
All mixes shall meet the performance requirements established by this section. Any proprietary material intended for supply must be tested against the requirements of this specification and no material shall be installed based on claims or representations made by the manufacturer.

Fig. 3.6. An artist's impression of One Central Park, Broadway with its planter boxes on each level of the apartment towers. (Source: <http://www.centralparksydney.com/>.)

Specification 1. The compost product used to amend the site topsoil shall have been tested for compliance to the requirements of AS 4454:2003 for a composted soil conditioner and approved before the transport of material to the blending site.

Specification 2. The blended mix shall comply with AS 3743(2003) Potting mixes for premium general purpose requirements.

Fig. 3.7. The installation and planting of the planters situated at each level of the high-rise building. (Source with thanks to: Michael Ball, Design Landscapes.)

Table 1. Chemical performance specification for the Soil type A growing medium

Item		Units	Horizontal planter media
Cation analysis	Sodium 3	%ESP*	< 5
	Potassium 3	%ECEC	3–11
	Calcium 3	%ECEC	65–80
	Magnesium 3	%ECEC	12–20
	Ca:Mg	Ratio	3–8
	Aluminium 3	%ECEC	< 1
Organic matter 7		% by mass	< 25%
Man-made foreign matter (e.g. glass, plastic, etc.)		% by dry weight	< 1

* ESP is Exchangeable Sodium Percentage, which measures the proportion of cation exchange sites occupied by sodium. Soils are considered sodic when the ESP is > 6 and highly sodic when the ESP is >15.

Methods: Mehlich 3: Rayment and Lyons (2011), Walkley-Black or dry combustion method. Performance measures of each soil type after amendment. Testing is to be conducted on amended material before placement to determine fertiliser type and application rate to achieve the soil chemistry listed in this table. It is not acceptable for fertiliser(s) and other amendments to be applied as recommended by an approved laboratory after soil placement to bring the chemistry of the material into specification.

Drainage sand
General description

Specified as a silt-free non-calcareous silica-type sand of medium grain size, free of silt, clay and organic matter. The sand is used to protect the geotextile and drainage layer against clogging by trapping silt and clay.

Performance specification (compulsory)

Particle size range	Units	Requirement
> 2 mm	% w/w	< 5
1–2 mm	% w/w	< 10
0.5–1.0 mm	% w/w	20–40
0.25–0.5 mm	% w/w	30–50
0.1–0.25 mm	% w/w	20–30
< 0.1 mm	% w/w	< 5
Organic Matter Content (AS4419)	%w/w	< 1
pH 1:5	pH units	< 6.5

Apply on top of the geotextile fabric a 50 mm layer of sand where soil depth is greater than 500 mm. Use 30 mm sand layer in soil installations with a total depth of < 300 mm.

Section 3: Quality control (QC)
Type testing

Before accepting any soil material or supply bid, laboratory testing is to be conducted to confirm compliance of the soil with specifications. Such testing may show non-compliance and a need to amend the material and re-test until compliance is achieved. Allow sufficient time for a process of testing, agronomic advice and re-testing to occur, generally 4 weeks before expected placement date.

All testing as required by the specification shall be arranged and carried out by the Contractor and all test results records made available to the Client. The cost of all such testing shall be borne by the Contractor.

The recommended soil testing laboratory for this project is:

SESL Australia Pty Limited (SESL)

16 Chilvers Road, THORNLEIGH NSW, Australia 2120 Tel: +61 2 9980 6554, Fax: +61 2 9484 2427

The Contractor is then to adjust the proposed soil mixes in accordance with the agronomists report supplied for each material by SESL and continue re-testing until compliance is confirmed by test reports.

Ongoing QA/QC Procedures

The minimum frequency of testing for ongoing QC purposes shall be 2 at approval (Type Test certification) then 1 per 50 m^3 or part thereof. Variations to this suggested frequency are permitted on the submission to the Principal's Representative of an alternative testing program that clearly achieves the desired outcome of quality control. Materials supplied from operations that have a third party endorsed Quality Assurance Program may be acceptable pending submission of the relevant documentation. Approval of any such Quality Management System remains with the client or their representatives. In the case where the Contractor has a system in place, provide a written description of the quality management system.

Sampling uncertainties can result in inaccurate test results. True quality control testing can be achieved by developing a standard method based on sound principles, consistently using the standard method, and documenting any adjustments caused by a typical conditions at the time of sampling.

The following sampling procedure has been included as a guide:

a) Remove and discard the surface material to a depth of 200 mm. If necessary, use a shield to prevent loose particles from moving into the sampling area. Samples are best collected from ~1 m up vertically from the base of a stockpile.
b) Remove sufficient material, using a scoop or shovel, from each location to constitute the sample increment, taking care to avoid spillage and to avoid contamination from the ground.
c) Sampling should produce significantly more material than is required for testing. The total volume of the sample must then be mixed and reduced to quantities appropriate for testing (i.e. ~2 kg for standard soil testing).
d) The process of dividing the composite sample must be done carefully to ensure that subsamples are uniform and representative. Conduct sample division using a standard coning and quartering method:

- All grab samples are placed on a plastic sheet and thoroughly mixed
- After mixing, the corners of the sheet are lifted simultaneously, thus causing the sample to mound in the centre of the sheet
- The sample is divided into quarters, and two opposing quarters are removed and discarded. This process is repeated until the sample is approximately twice as large as needed
- The last two quarters removed should be set aside for storage (library samples) and not discarded.

The two (2) identical samples should be placed in clean bags or containers and clearly labelled with the following information:

- date of sampling
- job reference
- type of material (i.e. specification)
- location of stockpile
- amount of material represented
- name of sampling operator.

The sample for testing should then be forwarded to an approved laboratory for testing for compliance to the specification and the library sample set aside in a safe place for future reference if required.

Non-compliance

Non-compliant material is that material brought to site by the Contractor that fails to comply with the performance specifications listed in **Specifications 1, 2** and Table 1 due to the practicalities of available materials or in some cases, a desire to use recycled materials.

Under some circumstances minor non-compliance may be acceptable and ultimately be approved for use if sufficient documentation and supporting laboratory testing from a suitably qualified agronomist is submitted, stating that the proposed material is equivalent or superior to the material specified or that the non-compliance does not compromise fitness for purpose or a practical means of correction (e.g. use of the liquid feed system) is available to correct the installed mixes.

No such material may be accepted unless a written statement of acceptability from a qualified and experienced landscape agronomist is provided.

All costs associated with testing, reworking, removal or replacing any material that the Principal's Representative deems to be unsuitable for construction because of non-compliance to the performance specification shall be borne by the Contractor.

Records

The Contractor shall keep and maintain all Quality System records as required by AS/NZS ISO 9001 and this Specification. They shall be systematically recorded, indexed and filed so as to be retrievable and accessible to the Principal's Representative or an appointed Quality Auditor on a job basis within one working day of requisition.

Quality register

Conformance records shall be stored and maintained such that they are readily retrievable and in facilities that provide a suitable environment to minimise deterioration or damage and to prevent loss.

Storage

The Contractor shall make the quality records available to the Principal's Representative at all reasonable times. If requested by the Principal's Representative, the Contractor shall provide copies of the records or test results at no cost to the Principal.

One critical aspect identified by the horticulturist/soil scientist was light distribution. Large multi-tower building with deep foyers and shadowing from within and surrounding buildings often present undiagnosed light problems. The engineers conducted light/shadow analysis (Fig. 3.8). These were then integrated with wind and exposure diagrams and simplified into an overall light/exposure diagram (Fig. 3.9). This

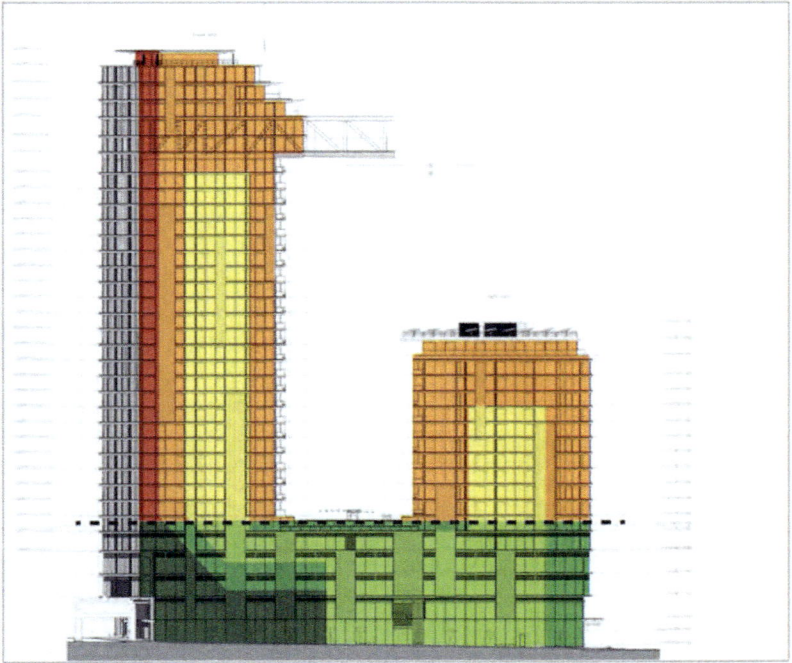

Fig. 3.8. Overall light/wind exposure diagram by Aspect/Oculus. (Source: with thanks to Aspect/Oculus.)

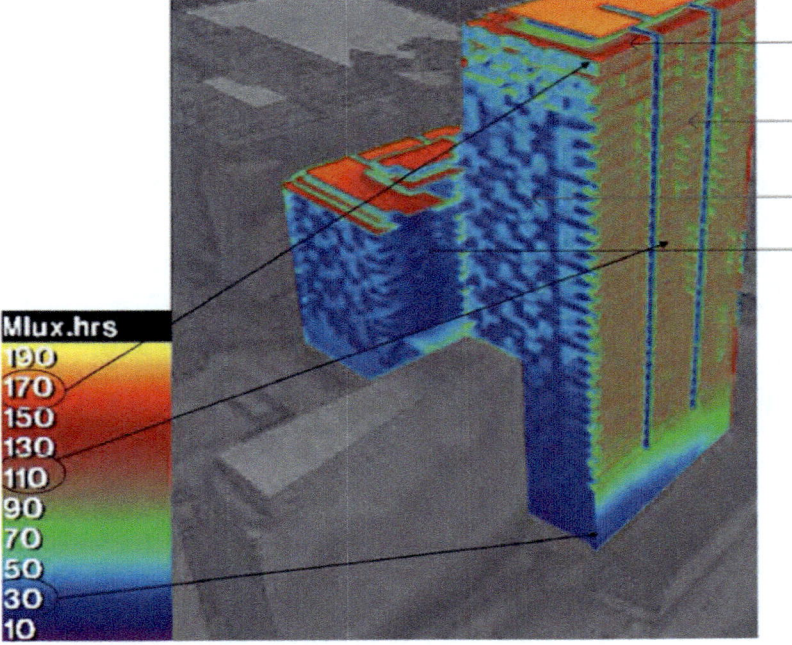

ONE CENTRAL PARK FROM THE SOUTH

Fig. 3.9. Light distribution study for One Central Park from the south. (Source: with thanks to Turf Design Studio.)

Fig. 3.10. The installation of the eastern facade 10 years after installation.

highlighted just how low the light levels were on the south-east of the lower towers and entrance foyer. The studies also then gave a logic to the plant selection, which was blocked up into zones based on light and exposure tolerance.

Plants tolerant of very low light levels were chosen for the south-east lower and foyer zones and full sun-, heat- and wind-tolerant plants for the upper north and north-west sides. As a result no low-light issues occurred; this is not the case with other buildings we have had to diagnose after the fact. Refer to Fig. 3.10.

3.16 LESSONS FROM FAILED PROJECTS (THREE EXAMPLES OF COMMON LANDSCAPE SOIL FAILURES)

Many good landscape design ideas fail, or do not perform well in practice, because of the use of inappropriate or unsuitable soil or growing media. The use of unsuitable landscape soils occurs because of a four-fold breakdown in the contract system.

Contract breakdown 1: A lack of understanding by designers and contract specifiers on how soils should be specified. Here is a typical example of the flawed 'recipe-based' approach that is commonly used by landscape architects:

- 1 part Kenthurst Black Loam
- 1 part sand
- 1 part aged mushroom compost.

There are three major errors inherent in this specification that could lead to failure of the landscape:

1. It was made at a time when 'Kenthurst Black Loam' was no longer available, meaning the specification is impossible to supply and would immediately require a variation.
2. There is no indication of what pH or any other soil property is required. Mushroom compost is highly alkaline because it contains excess lime, so any plant intolerant of lime (most plants) would suffer chronic 'yellows' (iron deficiency chlorosis). Also, mushroom compost is often extremely high in phosphorus so phosphorus-sensitive plants (many, though not all, Australian natives) would die or suffer chronic phosphorus toxicity. The landscape in this case study specified many *Grevillea* and *Banksia* plants that are known to be highly phosphorus sensitive.
3. The specification makes no mention of physical properties, and the use of a fine sand combined with a structure-less loam is likely to provide severe drainage problems (see 'Failed project example 1' below).

Contract breakdown 2: Few commercial soil suppliers conducting any (or adequate) research or trials on their products.

In litigation cases, the court will deem that a soil supplier is expert in soils. Therefore, if a soil supplier supplied the above soil mix knowing there were many P-sensitive native plants to be planted in it, or if asked to supply a soil suitable for native plants, a court could make a finding against them because they should reasonably have known that 'mushroom compost' is highly likely to be toxic to these plants. Some commercial soil suppliers have staff with adequate qualifications and experience, but there are many smaller soil suppliers that do not.

Contract breakdown 3: The temptation by landscape contractors to source a very low cost, inferior product. This can often be client driven and not the fault of the contractor. This often occurs because checking, testing and quality assurance checks are removed from the contractor's scope (either by the client to save costs or the contractor to win the job on a lower contract sum). This can be referred to in the industry as 'value engineering'. Contractors often resent the cost, time, perceived imposition and the idea of an authority (landscape architect or soil scientist) controlling parts of their contractual responsibilities. This is evident mostly on smaller projects where contractors feel they would rather take the risk of non-compliance. As well, even where landscape contractors' intentions are honest, levels of training and experience vary widely and it is surprising how many would not know that mushroom compost is toxic to many common landscape plants.

Contract breakdown 4: A lack of, or inadequate project engagement of, the landscape architect or soil scientist (or other independent expert authority) during the installation process. Normally, they would, through their professional standards obligations, industry responsibilities and indemnities, ensure the soil specifications and soil validations are met. Often this process is managed by the builder, who in some projects could be deemed a conflicted party (by way of being self authorising). Often the certifier overlooks or does not put adequate attention on the landscape and soil components of the project.

In early 2024, bonded asbestos was found within organic mulch across multiple parks, schools, and private and public open spaces in Australia – reminding us of the importance of these validation and certification checks.

Make sure to include soil **Specifications G1–G3** (in Chapter 6) together with landscape documentation regarding compliance and validation.

3.16.1 Failed project example 1

Part of the project that employed the soil 'recipe-based' specification above called for its use on a rooftop turf installation. The idea was for a nice open area of lawn for workers

Fig. 3.11.   The two images are of buried organic matter creating an anaerobic perched watertable resulting in waterlogged turf soil and limited available soil depth.

to have lunch and relax during breaks. The specification called for a waterproof membrane over the concrete, a drainage 'cell' over that to conduct water away, a geotextile fabric over that then, very importantly, a layer of sand to prevent blockage of the geotextile with silt, then the soil layer and then turf. All this is good and 'to the textbook'. The problem was that the soil was too fine and created a 'perched' watertable because of the texture difference between the fine soil and the coarse sand. The black layer in the photos in Fig. 3.11 represents the totally waterlogged 'capillary fringe' where the sulphate from the mushroom compost (also typically high in gypsum or calcium sulphate) had been reduced to black sulphide with the accompanying smell of rotten egg gas and resultant failure of the landscape.

3.16.2 Failed project example 2

In a major redevelopment project, the landscape architecture firm went to the extent of including a performance-based specification similar to those advocated here for, among other things, the installation of a rare and endangered native (Australian) plant community. Critical was the very low phosphorus level required by the specification. The contractor brought in the soil–compost mix recommended for natives by their local supplier. The supplier had clearly not allowed for the provision of analysis and, when asked to provide one before installation, the supplier provided the following information:

- Compost mix of equal parts mushroom, chicken and duck manure
- Compost analysis: 1200 mg/kg of 'available' phosphorus.
- 1 part in 5 with site soil.

The compost was composed of equal parts mushroom compost, chicken manure and duck litter. Note the mix shows 1200 mg/kg of 'available' phosphorus. The error, which would have been fatal to many of the specified plant list species, was picked up, not by the landscape contractor or the soil supplier, but only by the last-minute insistence on adherence to the specification by the landscape architect.

3.16.3 Failed project example 3

A local council had the laudable idea of showcasing some of its original native floral types in a waterfront park. There are no records of any soil specification for the project or even of who supplied the soil. After multiple plant losses and deaths and three re-plantings over 4 years, some soil tests were finally commissioned. The problem shown in Fig. 3.12 is severe manganese deficiency induced by very high pH and phosphorus levels. It was diagnosed by foliage analysis that showed only 1 mg/kg of Mn in the foliage. The soil was very high in lime, phosphorus and organic matter, all of which conspire to reduce Mn uptake. It was totally unsuitable for the purpose of growing native vegetation. The problem could not be rectified and all the soil had to be replaced. The council gave up and reverted to lime-tolerant plants only. A soil investigation and technical performance specification as advocated in this book would have saved tens of thousands of dollars in public money.

Fig. 3.12. Severe lime-induced manganese deficiency in *Pomaderris ferruginea*.

4

Soil design

4.1 PROJECT INITIATION

A project is almost universally initiated by a client and often variously influenced by their expectation or pre-conceived notions of what they want. A designer's job is not just to prescribe a solution, but to lead the client to see the logic of their design process through a thorough approach to:

- landscape objectives (see **Appendix B**, Section B1)
- site soil analysis (see **Appendix B**, Section B2)
- articulation of design principles and solutions (see example below).

Very often edaphic conditions are well down the list of considerations, taking second place to the novel, the showy or the rigidity of formal layouts.

A *classic example* is a hedge or avenue of a single species of plant crossing through differing soil conditions and gradients. The avenue will not grow uniformly and, at worst, if waterlogged for example, a whole section may die if the soil is not artificially changed to accommodate the hedge (Fig. 2.1).

The *design solution* is to either:

- change the soil, or
- change the design to cope with the edaphology.

Edaphic conditions influence design more than is commonly appreciated and site soil analysis needs to form a part of the planning from a very early stage, preferably before concept design and certainly before detailed design.

It is important to note that the presence of any soil-borne disease or fungus influences the soil and therefore the soil treatment, design, species and specifications.

4.2 SITE SOIL ANALYSIS

Assessment of the site soils should be a normal part of site analysis and is essential in working out the best design approach, which is one that is both cost effective and environmentally sound. In Fig. 4.1, the soil assessment would be advisable around the existing pool, especially given the failing cypress hedge next to it. Pool wash contaminating the soil with excessive salt or chlorine is an obvious culprit, as well as a soil volume issue as seen in Fig. 4.1 (right) where a concrete base was installed under synthetic turf. A quick on-site pH test could point to other soil toxicities or soil deficiencies, but a proper soil analysis back in the laboratory is also recommended for completeness and to improve on the guesswork, providing you and your client with confidence and justification in paying for any rectifications.

Fig. 4.1. Localised out of balance pH caused in part by salt spill from the pool and reduction in soil volume have caused dieback in this cypress hedge.

Only once there is some knowledge of the soil resources on site, or a consideration of the soil resources required to successfully implement a design, should any concept or detailed design be finalised.

For small or low budget projects, Table B1 in **Appendix B** may be a suitable checklist for characterisation of the soil without the engagement of a soil scientist.

Often in urban situations, there may be no or limited access to the soil at the concept phase of a project (e.g. the area is taken up by a building or hardstand such as a sealed carpark). Until access to the soil is possible, presumptions need to be made based on area soil maps and probable previous construction techniques on the site. Soil specifications should set out the method for investigation at the demolition phase and a potential variation statement, similar to that of excavation works, and tenderers should provide a similar cost variation statement.

Understanding the soil resource

Understanding the soil resource is as equally important and influential a step in the process of a well-considered design approach as any other site attribute, such as slope, aspect or context.

Methods for specifying the minimum standard of site soils analysis required are given as **Specifications A1** and **A2** and are summarised in Table 3.9 (see also Chapter 3 and 6).

Table 3.9 outlines general characteristics and observations required in the site analysis phase. This table may be used as a checklist for experienced industry professionals or may be used as a 'scope of works' list for urban landscape soil scientists for field analysis.

4.3 LANDSCAPE BALANCE (PRIORITISING PROJECT FACTORS)

It is the landscape architect's responsibility to holistically consider and balance the priority and weighting of the set objectives and methods outlined in this handbook including:

1. the *site investigation:* carrying out proper site analysis including soil and landscape analysis
2. the *Soil Approach Method:* determining the appropriate method, such as importing new soil, conditioning the soil, rehabilitating or using the existing site soil together with the type of soil profiles to be installed
3. finalising the *landscape objectives* for the project and disseminating the objectives via communication and collaboration with the client and the project team.

These three key considerations come together to influence the landscape design and the landscape specification, which, in turn, generates the landscape construction process.

This typical process is further outlined in Section 4.7 'The Soil Approach Method'.

4.4 LANDSCAPE OBJECTIVES

Working out what the designer and the client want to achieve sounds very fundamental and is, for most designers, intrinsic to how they practise, but it is also a key step in achieving a high-quality outcome. Cost-effective and environmentally sound design intent can be easily carried through the design process with some simple considerations to the design approach, with the first step being to establish the landscape objectives.

The landscape factors and objectives should be written down in order of priority by both the designer and client to provide a constant reference checklist to keep the design process focused.

Design objectives often contain the following considerations:

- authority requirements
- budget
- environmental considerations
- timing of construction
- time allowed for landscape maturity
- intended maintenance
- longevity, expected lifespan of the project (landscape)
- aesthetic qualities
- landscape function and purpose
- amenity and management
- value over time
- site factors, including the influence of existing site soil or soil importation requirements on design choices.

4.5 INFLUENCE OF SOIL ON DESIGN

The nature of the soil on a site can have just as much influence on design as any other site factor. An obvious example would be very shallow artificial soils on a landfill that really precludes the adequate growth and development of large trees.

The example of the Barangaroo development in Section 3.13 is one where topographic position, aspect and slope influenced the choice of vegetation type, which then determined soil depth and choice of fertility level.

Soil factors also influence the landscape objectives in other ways, so early analysis of soil factors as part of overall site analysis is important to uncover issues that may imbalance or even rule out some design objectives. An obvious example would be the presence of alkaline soil, which rules out a great many plant species that are intolerant of such soils (refer to Table 3.3).

Another typical example is where *environmental sustainability* and *cost effectiveness* are of high priority and the landscape design calls for completely new soil to be imported in order to sustain particular plant species not suited to the site conditions. This creates an imbalance in that the landscape objectives are not being met, particularly where carbon emissions are being measured as part of the project objectives.

A more appropriate effective alternative solution in this instance would be to use site soil, if deemed appropriate, and change the types of plants to those that are suited to the existing soil conditions. Further, some topsoil may be stripped and stockpiled, and cleared vegetation may be chipped and reused on site as mulch, provided the mulch is well composted (at least 6 months matured). There may also be an opportunity to improve the stockpiled topsoil on site through planting a 'green manure' or through phytoremediation.

Another example that often occurs is where site soil has been fertilised with phosphorus and hence the use of P-sensitive native plants is precluded. Alternative plant palettes must be chosen or the soil must be completely replaced.

This adaptive approach:

- requires good site soil information before design concepts can be finalised
- saves money that can be redirected into feature landscape items such as mature trees and increased maintenance
- reduces long-term maintenance as landscapes are better adapted edaphically
- reduces the environmental impact of both purchasing and disposing of existing materials (fill material disposal)
- reduces the impact on quarried material (for purchased material)
- reduces transport costs (for soil removal and delivery)
- reduces the potential of importing contaminants such as weed seeds, recovered wastes or soil-borne pathogens in imported material.

Among the most extreme examples of soil influencing design are:

- salinity and its distribution on a site
- poor drainage and waterlogging and its distribution
- heavy clay soils that preclude certain vegetation types
- alkaline or acidic soils that restrict the plant list to tolerant species
- phosphorus-contaminated soils that preclude the use of P-sensitive plants.

Where there is a lack of suitable soil on a site, this constraint does not occur since soil can be designed and imported to suit any landscape aim, with budget considered.

4.6 SOIL CHOICE VERSUS DESIGN

There are many other examples of where it is possible to 'design out' an issue or problem, as opposed to 're-engineering' the existing conditions just to suit an unsuitable or poorly considered design.

It should always be a priority to 'design out' issues. After all, effective landscape architectural design is influenced by site (including soil) and context as well as budget and client aspirations. During the design process, the designer should create opportunities to:

- create a solution 'by designing it out', rather than making technical, engineered changes and additions to suit a design. This might be choosing native plants tolerant of acid low-fertility soils rather than conducting intensive soil improvement works.
- take balanced approaches that refer back to the prioritised landscape objectives to facilitate the choice between the two above approaches. Ideally, site soil

information would be available before concept design is finalised and certainly before detailed design is complete.

Often, designing out a problem or solving an issue through design will be more cost effective and a more environmentally sustainable choice (and is a sound way to illustrate the positives of this approach to clients).

A classic and common example of imbalance between design and technology is where a building is designed for green roof installations without consultation with a soil technologist. Often, insufficient consideration of soil weight occurs and the soil technologist is expected to design soils of impossibly low density or a density that is very expensive to achieve. An extra 25 mm thickness of concrete, for example, would have allowed more normal, widespread and cost-effective, light-weight soil to be installed.

4.7 THE SOIL APPROACH METHOD

Decisions about what site soil can be reused and what and how much soil or soil improvers need to be imported can only be made after site soil analysis has been performed and the report describing the soil assets is available. Once this is done, the Soil Approach Method or combination of approaches from Table 4.1 and Fig. 4.2 can be chosen.

At this stage, there is a strong interaction with the design process, client and team; that is, the soil approach will influence the landscape design approach and vice-versa. Input from the soil technologist to finalise the soil approach designs will be important at this stage.

Increasingly, informed clients and developers are becoming more aware of site soil as a site asset and are keen to balance improved sustainability objectives and cost savings with a well-thought-through Soil Approach Method, versus the traditional approach of scalp and dispose or degrade the site soil resource and pay to import landscape soil at the end of the project.

Obviously, with most development sites being 'infill' developments, and less available space on sites for stockpiling or fencing off (protecting), this balance of project objectives with sustainable factors is a key part of the design process to coordinate with the client and project team.

The prerequisite to working out the soil approach is the site soil survey and report, which will help the landscape architect to formulate and present the design options to their client and project team. Once a draft concept design is created, consultation with the soil technologist will result in a fit-for-purpose Soil Approach Method to sustain that design.

4.7.1 Towards not just sustainable, but regenerative soil environments

The Soil Approach Method should also take into account a hierarchy considering carbon emissions, sustainability practices and environmental performances that can be utilised in approaches to rating and measurement tools. Such tools as as 'Towards Absolute Zero', 'Transform to Net Zero', 'GHG Protocol', 'Scope 1, 2, and 3 Carbon Emissions', 'Biodiversity Design Guides' and 'Green Star Rating', 'National Australian Built Environment Rating System' (NABHERS) and 'International Sustainability Rating System' (ISRS) performance-rating systems are increasingly becoming compulsory during the planning process. Refer to Table 4.1.

A suggested conceptual hierarchy is given in Fig. 4.2.

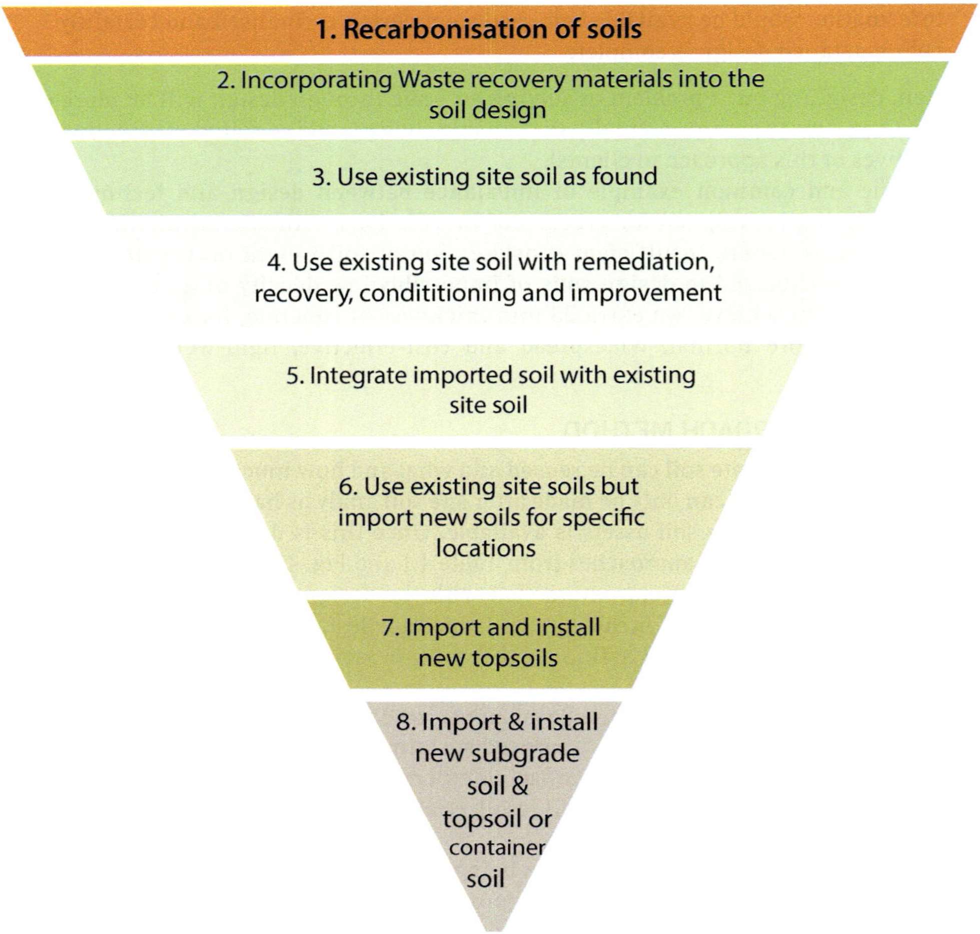

Fig. 4.2. The Soil Approach Method sustainability and carbon footprint hierarchy.

Table 4.1.   The Soil Approach Method chart

Action and preference	Emissions	Description of emissions	Notes
Consider landscape design with regard to ongoing management, upkeep and maintenance.			
1. Recarbonisation of soils	Negative emissions	Site derived materials such as composts made from site resources.	Minimal landscape approach. Consider landscape design with regard to ongoing management/upkeep/ maintenance.
2. Incorporate waste recovery materials into the soil design (from either on-site or nearby off-site).	Zero or negative emissions: high priority	Recovery soil improvements are imported. Note any added transport to the site and any on-site fuel emissions at installation.	The more local the waste product is located to the project site, the less energy input, if the blending and stockpiling occur on site. Usable waste product components can include excavated material, manures, mulches and green waste composts.

3.	Use existing site soil as found.	Zero to very low emissions	Consideration of exclusion fencing to protect soil structure and soil micro-biome during construction works.	Investigate site soil to determine if it is appropriate, and if the nutrition is adequate for optimal growth and health of the proposed plant material. Most common in land rehabilitation and mass planting projects.
4.	Use existing site soil with remediation, recovery, conditioning and improvement.	Very low emissions	Stripping and stockpiling topsoil and subsoil. Consideration of the imported and embodied energy of soil additives. Types include: natural/organic to chemically derived.	Recovery and reuse may require imported materials such as ameliorants (e.g. lime and gypsum), fertilisers and integration of organic matter to the top horizon and physical decompaction and amelioration of subsoils. Consider both organic and synthetic fertilisers.
5.	Integrate imported soil with existing site soil. Items 5–7 are a lower priority but are sometimes unavoidable.	Very low to moderate emissions	Consider transport emissions (distance and quantity). Consider derivation of imported soil (mined, recycled/recovered waste).	Such as where depth needs to be increased or texture needs modifying (e.g. adding sand to playing fields or soil build up (fill) to increase levels).
6.	Use existing site soils but import new soils for specific locations.	Very low to moderate emissions	Using local suppliers is preferable.	A common specific area soil import may be a loamy sand for sports fields, and a potting media for raised vegetable gardens and rooftops. Cost benefits and environmental benefits accrue by specifying different soil approaches for different parts of the project.
7.	Import and install new topsoil, ameliorate subsoil or subgrade	Moderate to higher emissions	Consider export of usable natural resources off site if not usable on site. Consider transport emissions to site and within the site (i.e. location of stockpile). Consider soil supplier emissions.	Where the subgrade is adequate and there is no available topsoil (e.g. site has been stripped of all topsoil, such as in a contaminated area or brownfields site, or where space constraints mean stockpiling could not occur).
8.	Import and install new subsoil and topsoil or container soil media.	High emissions: lowest preference	Consider the transport and environmental impact of disposal of soil to the site (e.g. is the site soil going to landfill, is the disposal via 'capping' if it is contaminated, consider the possible methane emissions of organic content). If importing material, consider whether the imported materials may have come from finite resources (if mined or dredged from creek beds/ alluvial plains, greenfield sites or otherwise vegetated areas).	The most common example is where no existing soil is available, or where the landscape design requires a completely manufactured soil media (such as on a rooftop or vertical wall). This will be the case in most urban developments. Another example is installing a full soil profile of imported subsoils and topsoil layers over capped landfill or contaminated land. Consider the use of recycled soil from other nearby sites such as excavated material.

4.7.2 Recarbonisation of soils

The intent of this approach is to actively improve the existing site soils by increasing the soil carbon content through applying composts; this will enhance current and future vegetation growth and soil biota. Here are some examples of this approach:

- Utilise and reuse site derived material that may otherwise have been disposed off-site, such as a site made compost from vegetation clearance.
- Where this is not possible, choose recycled waste-derived composts form the local area.
- Consider long-term possibilities: that the site may generate usable soil improvement waste products that can be reused on site to further improve soil health and soil carbon. One example is restaurant or residential kitchen organic waste composting and use.
- Consider on-site water harvesting and capture, if necessary run through biofiltration, raingardens or wetland systems for on-site landscape recharge, and enhanced plant growth and carbon storage.

4.7.3 When can you use existing site soil as found and *in situ*?

While the use of natural soil landscapes is becoming rarer in projects, this can still occur such as:

- around trees to be retained and protected on sites
- areas with heritage values, in parklands
- botanic gardens, parks and arboreta
- rural estates
- environmentally sensitive developments, and
- natural vegetation tracts.

When using natural soils, it is still important to conduct a site soil resource survey so that the background soil fertility and physical properties can be assessed. Even so, some of the above listed landscapes being redeveloped may still be degraded by nutrient loss and erosion, and may need rectification, amelioration and improvement for a higher value use, depending on the landscape outcome.

When retaining and using natural soil *in situ*, the main considerations are:

- matching vegetation to available resources (e.g. availability of irrigation, budgets for fertilisers, imported organic matter and soil ameliorants)
- identifying areas of excessive drainage, waterlogging (poor drainage), impeded rooting depths and other factors likely to have an impact on the choice and performance of landscape plantings
- identifying improvements or amelioration of adverse conditions required to support the chosen landscape treatments. Improvements could include elevation of soil fertility or improvements to drainage of wet areas.

When using site soils **Specifications A1, Specification A2** and **B2** will probably also be required to investigate, improve or ameliorate the soil for the intended use. In the case of using site soils *in situ*, it is generally not possible to treat subsoils unless they are exposed, so the specifications for preparation of subgrades will therefore not be required, although it is recommended that subsoils be investigated (**Specification A2**) anyway, so that during tree planting where subsoil is exposed, subsoil may be ameliorated if required.

The main decision to be made will be what level of soil fertility is required; thus, **Specification B2** provides for a low, medium and high level of soil fertility to cover most landscaping needs.

4.7.4 Protecting the soil – *in situ*

There may be opportunities where it is appropriate for the protection of soil zones during site preparation works (demolition works) and construction works. This is typically done by fencing off landscape zones and is commonly done where trees are also to be protected.

By establishing exclusion fencing to landscape zones, this can:

- protect the soil structure and soil biota
- retain and not contaminate the soil fertility (such as by construction waste, e.g. concrete wash or paint wash)
- significantly reduce soil removal costs and landscape soil improvement works costs.

In the event of protecting the soil, considerations and design specifications by the landscape architect may include the following.

Physical protection/barriers: such as protective fencing and associated signage, protective mulch layers or designated paths with suitable ground protection such as temporary timber sheeting or rumble boards, suitable light irrigation to suppress dust and the effect of adjacent changes (which may significantly affect the microclimate), silt and dust barriers. Where fencing off is not an option, and albeit not perfect, in tight development sites, the location of site sheds (raised up on pad footings) and even setting aside a zone for the construction team's lunch zone (possibly with protective boards on the ground) can sometimes protect the landscape soil profiles better than if those zones were used as vehicle parking, access roads, materials storage or general construction.

Construction management: including suitable site inductions and restrictions on certain activities allowed around the soil protection zones. Generally this is difficult without physical barriers and signage.

Preventative measures: diversion, filtration or collection of construction liquids and other contaminants entering the soil protection zone to be controlled, such as contaminated site waste water. This can be achieved by utilising:

- erosion control measures (coir logs, hay bales, sand bags)
- installing silt fencing
- contour mounding (swales and berms)
- temporary suitable hydroseeding or revegetating before construction.

4.7.5 Reconstructed soils, recovered or imported soil source

Regardless of whether soil resources are recycled from site or imported, it is necessary to think about what type of soil profile will be required to support the intended landscape treatment in a sustainable way; that is, with a minimum of maintenance inputs (such as irrigation and fertiliser) and with reasonable growth and appearance.

4.8 CHOOSING THE PROFILE FORM FOR PART I OF THE SOIL SPECIFICATIONS

Table 4.2 outlines each of the four soil profile horizons. The naming of the soil horizons is derived from soil horizons found in natural soils. This table shows the comparison of natural soil horizons to constructed soil horizons.

Table 4.2. Comparison of natural and reconstructed soil horizons

Natural soil designation	Reconstructed equivalent
O Horizon	Mulch
A Horizon	Topsoil
B Horizon	Subsoil
C Horizon	Subgrade*

*Note that subgrade is not generally considered part of the functional soil profile. It may be landfill capping or mixed compacted cut or fill surface.

4.8.1 Specifying a two layered profile (or an A/C profile)

An *A/C profile* (Table 4.3) is one where topsoil is placed directly over an ameliorated or unameliorated subgrade and there is no effective subsoil to provide significant rooting depth. It is suitable for shallow-rooted annual and perennial vegetation with simple structure such as:

- annual herbaceous species or mixed meadow
- grassing and turf including lower grade sports fields
- groundcovers with small shrubs or tough small trees.

Disadvantages of this profile form are:

- potentially shallow rooting depth especially with clay or rocky subgrade which could impact tree stability
- tendency to waterlogging where subgrade is impermeable.

Table 4.5 summarises the soil horizons needed to be installed and which type of profile will suit the design objectives. Generally, it is desirable to rebuild a fully developed soil profile with:

- topsoil (A)
- subsoil (B) and
- subgrade (C) horizons.

An A/C profile is not considered adequate for any tree or forest vegetation, unless using very compact, drought and waterlogging-resistant species of low stature (mature height less than 5 m) or low-quality mass planting such as on mine or rehabilitation sites, where stunting of the vegetation is acceptable.

Drainage may be improved by sloping, cambering or, if necessary, the installation of systematic drainage.

Table 4.3. A/C profile characteristics

Horizon	General properties	Function
A – Topsoil	Good permeability (friable or granular) Organically enriched Appropriate nutrient reserves Biologically active	Nutrient reserve Root anchorage (part) Good aeration Fast water entry Moisture holding
C – Subgrade	No or limited root entry Firm incompressible base Permeable or impermeable	Physical support Drought moisture reserve Drainage layer (may be combined with artificial drainage)

Table 4.4.   Characteristics of A/B/C profile

Horizon	General properties	Function
A – Topsoil	Good permeability (friable or granular) Organically enriched Appropriate nutrient reserves Biologically active	Nutrient reserve Root anchorage (part) Good aeration Fast water entry Moisture holding
B – Subsoil	Adequate permeability (good structure) Low organic matter Balanced cation exchange Good moisture holding	Firm root anchorage Adequate aeration Adequate water entry Good moisture holding
C – Subgrade	No or little root entry Firm incompressible base Permeable or impermeable May be concrete on built structures	Physical support Drought moisture reserve Drainage layer Combine with artificial drainage

Table 4.5.   Summary for installing soil horizons (the profile form)

Profile type	Description	Recommended for	Unsuitable for
O	Generally called 'mulch', an O horizon is usually organic matter such as woodchip or bark but may be pebble or sand	All soils should ideally be mulched to reduce water loss from soil surface	Massed bedding plants and turf. Mulch that floats is unsuitable in bioretention zones, swales, raingardens, detention zones and the like
A/C	Topsoil is placed right over essentially unameliorated subgrade fill	Low-cost rehabilitation with shallow rooted grasses and groundcovers or low shrub layers. Can be used for sports fields with cambering and/or drainage	Any kind of tree or vegetation over 5 m. Vegetation intolerant of drought and waterlogging
A/B/C	A layer of improved subsoil (B) horizon is placed over unameliorated fill before placing topsoil. Where possible, subgrade is ameliorated to form the subsoil (B) layer	Minimum standard for garden, tree and shrub planting. Horizon depths should be increased as mature vegetation height increases	Cost of improving subgrade or importing subsoil may not be justified for large scale projects or indigenous species rehabilitation

4.8.2 Specifying a three-layered profile (or an A/B/C profile)

A mature natural soil profile shows three more or less distinct layers, or four if we include the litter (organic matter or mulch) layer.

A landscape soil can be reconstructed in an analogous way as illustrated in Table 4.4. *A/B/C profiles* are typically suitable for:

- perennial woody vegetation with complex structure
- groundcover, shrub layer and trees
- tree avenues and individual tree specimens
- parks with trees
- higher grade land rehabilitation
- shrub and tree borders and mass planting.

The nature of the three soil horizons and their behaviour and function is summarised in Table 4.4.

The only disadvantage of constructing a more complex soil profile like this is increased initial cost.

The subsoil layer, on top of subgrade, may be manufactured from:

- some reserved natural subsoil from the original site soil
- site topsoil with no added organic matter
- ameliorated subgrade.

Subgrade may sometimes be ameliorated or improved to form a subsoil layer by:

- application of lime, gypsum or acidifying agents as required
- working to 300 mm minimum (e.g. with chisel ploughs or rippers) to reduce density, key and roughen the surface
- picking of rock, timber, masonry and rubbish
- fairing and shaping for, for instance, camber slope.

4.9 SPECIFYING SOIL DEPTHS

The performance specifications in Part II B4 provide a manner of specifying profile form and horizon depths for various vegetation types. These are recommendations based on experience and represent an optimum for good growth and development of vegetation with minimal maintenance or irrigation.

The recommended depths may be subject to variation at the discretion of the designer, depending on budgetary and soil availability limitations.

> Put simply, the greater the effective root depth (up to possibly 1500 mm) the higher the status of vegetation it can support with minimal maintenance.

4.10 SPECIFYING DRAINAGE TYPES

Although the key focus of this handbook is the soil specifications, and it is not our intention to include drainage specifications, it is essential that the principles of drainage be understood and drainage issues properly anticipated in the design approach and soil approach (see section 4.7, Fig. 4.2 and Table 4.1).

The presence of layers of fill or soil at depth that impede the downward movement of water can severely limit growth and development of vegetation, and can result in wholesale death of plants, either shortly after transplanting or even of mature vegetation following prolonged wet periods.

The problem occurs because roots are aerobic tissues and can be killed by even short periods of hypoxia (low oxygen levels) or anaerobia (zero oxygen) as shown in Fig. 3.11. Water-filled pores conduct oxygen by diffusion so slowly that they cannot keep up the supply to the respiring root.

Some plants, such as rice, aquatic and wetland plants and species growing in flooded landscapes such as many melaleucas, can develop special root structures (aerenchyma) that allow oxygen to diffuse down the root itself, but most species do not and roots quickly die if soil becomes hypoxic.

It can particularly be a problem when transplanting large tree specimens from nurseries, where deep narrow-shaped root balls are planted into soil, trapping the deeper roots away from surface oxygenated layer and into lower layers that can quickly become hypoxic if waterlogged by rain or over-irrigation. Plant loss shortly after transplant can most often be attributed, if not to drought, then to hypoxia in the lower root ball.

The other common situations that require drainage are heavily used turf with an A/C profile over impeding clay layers, and plantings at 'break of slope' (i.e. where a steeper slope changes to a shallower slope or flat area and water run-off from upslope keeps the area waterlogged). This latter problem also happens where water runs off hard surfaces onto garden beds or individual plantings.

It is very important to understand that subsoil drainage not only allows excess water to exit the profile, but also for air to enter, greatly increasing the depth of the aerobic zone and hence effective rooting depth.

Anticipating drainage problems should occur at the point of bringing the *soil selection process* (Section 3.9) together with the Soil Approach Method (Section 4.7). Carefully examine the design where any of the following soil types and land features occur:

- The underlying geology is a fine sediment or alluvium (i.e. shales and silty clay alluviums, such as river, lake and estuarine sediments).
- The natural soil is hydromorphic, as seen by the presence of peaty accumulations and waterlogged morphology.
- The location is low lying or on a lower slope.
- The ground relief is internally drained (i.e. very flat or even concave).
- The subsoil is clay or poorly structured clay loam or silty clay loam.
- The subgrade is compacted mixed fill likely to drain poorly.
- The installation is over a concrete slab or an engineered surface.

Design features likely to be adversely affected if systematic drainage is not installed are:

- heavily used sports and recreational turf – see interception and vertical drainage, Figs. 4.3 and 4.6
- break-of-slope plantings – see preventative and buried drainage, Figs. 4.3 and 4.5
- impermeable subsoil or subgrade, any planting type – see buried drainage, Fig. 4.5
- large (over 45 L install pot size), advanced and super-advanced tree plantings with impermeable subsoil or subgrade) – see interception drainage, Fig. 4.4
- at slope interceptions – see buried drainage, Fig. 4.5 (e.g. walls, kerbs, terraces running across slope and likely barriers to water movement creating 'dams').

4.10.1 Drainage types

Drainage can be categorised functionally as 'preventative' where steps are taken to reduce the amount of water getting onto, or remaining on, a site or 'curative' where, once infiltrated, the excess water has artificially installed channels through which to flow more quickly out of the profile.

Category 1: Preventative drainage

Surface drainage

Surface drainage is used to divert and prevent excess run-on from upslope from impacting a site or, where it does, make sure it gets away quickly and doesn't pond. The former is achieved by installation of diversion ditches and swales to divert and conduct water around the site (Fig. 4.3). The latter is improved by correct sloping to allow run-off to leave more rapidly and not pond on the site. It is also important to avoid obstacles to run-off.

Interception drainage

Interception drainage not only involves the interception of surface water flow but also includes the interception of lateral groundwater flow to prevent groundwater moving into a site and to lower watertables (Fig. 4.4).

Category 2: Curative drainage

Buried drainage

Sometimes called systematic subsoil drainage, we use the term buried drainage to encompass several techniques to keep both topsoil and the entire profile free from saturation (Fig. 4.5).

Vertical drainage

In landscape construction, the most commonly seen use of vertical drainage is to make vertical sand slits and drains in sports fields, but the technique certainly has wider applications (Fig. 4.6).

Cellular drainage

Of particular application on slab and built structures, products such as Atlantis® or Elmich® drainage cells or layers of drainage gravel are used (Fig. 4.7).

Filters

Filter media (Fig. 4.8) are used to prevent silt and clay from blocking drains and to stop soil falling into drains over time, which is very common. As well, filter media (designed correctly with specific plant material) will filter various heavy metals, toxins and soluble nutrients (beneficial and otherwise) and largely prevent them from entering into the water system. This may be achieved by physically trapping larger particles. Such soil particles can hold toxins, chemicals and heavy metals filtering them out of the water stream. Once residing in a filter a biofilm can develop metabolising organic contaminants or biodegrading them. Plants growing in biofilters can also remediate some organic contaminants. The use of geotextile or 'filter' fabric to line and wrap drainage systems can be problematic because silt and clay can block the cloth. Installing a layer of sand over the geotextile is strongly advised.

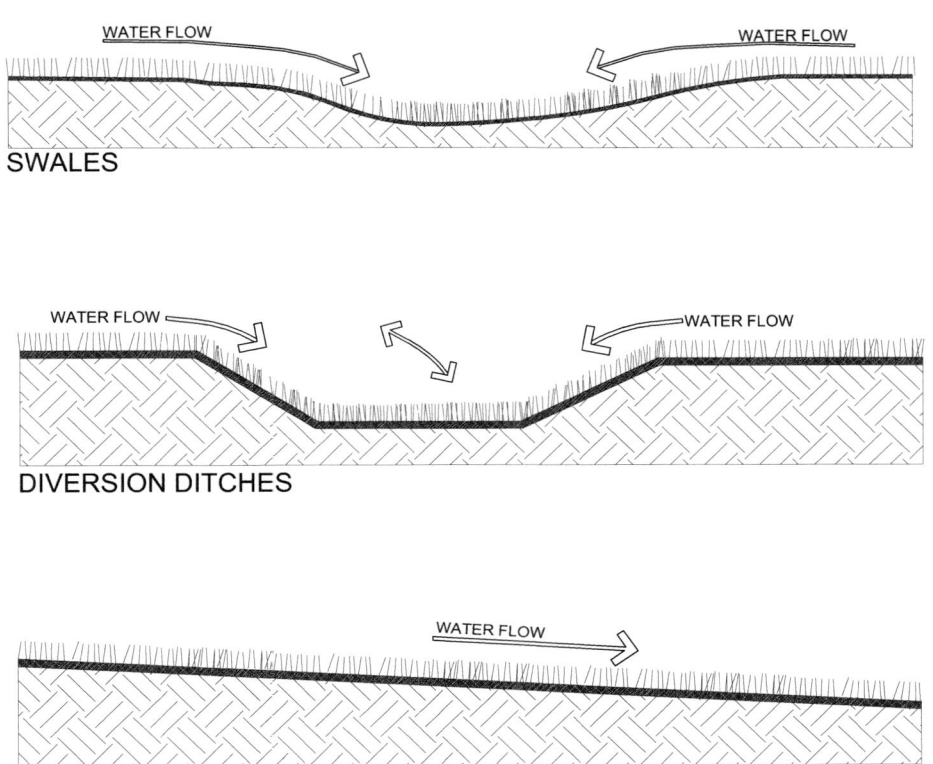

SWALES

DIVERSION DITCHES

SLOPE.

1:50 SLOPE

(A) PREVENTATIVE SURFACE DRAINAGE
NOT TO SCALE. DIAGRAMMATIC ONLY.

Fig. 4.3. Preventative surface drainage systems (from top to bottom): swales, diversion ditches and slopes.

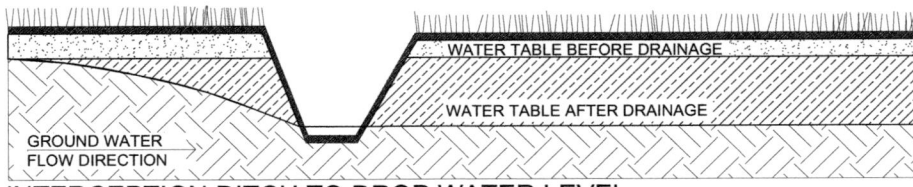

INTERCEPTION DITCH TO DROP WATER LEVEL.

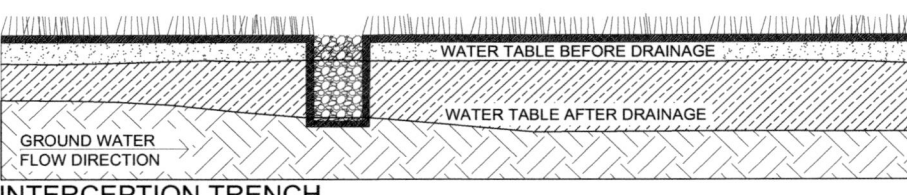

INTERCEPTION PIPE

INTERCEPTION TRENCH

INTERCEPTION DRAINAGE

(B) NOT TO SCALE. DIAGRAMMATIC ONLY.

Fig. 4.4.   Curative drainage types with interception of surface water flow (from top to bottom): ditch, pipe and trench.

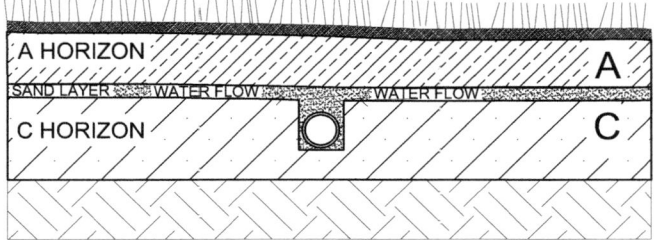

1. BURIED TOPSOIL DRAINAGE
RELIEVING PERCHED WATER TABLE BELOW A HORIZON WITH SAND LAYER

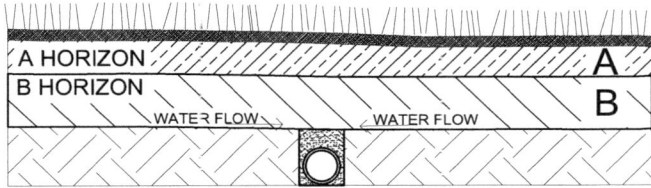

2. BURIED SUBSOIL DRAINAGE
RELIEVING PERCHED WATER TABLES.

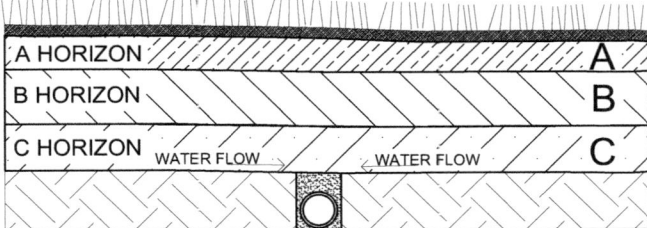

3. BURIED SUBGRADE DRAINAGE
RELIEVING PERCHED WATER TABLES BELOW THE B HORIZON

C BURIED DRAINAGE
OR SYSTEMATIC SUBSOIL DRAINAGE
NOT TO SCALE. DIAGRAMMATIC ONLY.

Fig. 4.5. Buried drainage types (from top to bottom): buried topsoil, buried subsoil, buried subgrade.

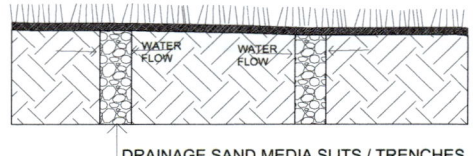

DRAINAGE SAND MEDIA SLITS / TRENCHES

VERTICAL DRAINAGE
COMMON IN SPORTSFIELDS BUT APPLICABLE TO OTHER APPLICATIONS.
NOTE: PROFILE HORIZONS NOT SHOWN.

(D) ## VERTICAL SAND SILT OR TRENCH DRAINS
NOT TO SCALE. DIAGRAMMATIC ONLY.

Fig. 4.6. A vertical drainage system (also known as sand slit or trench drains).

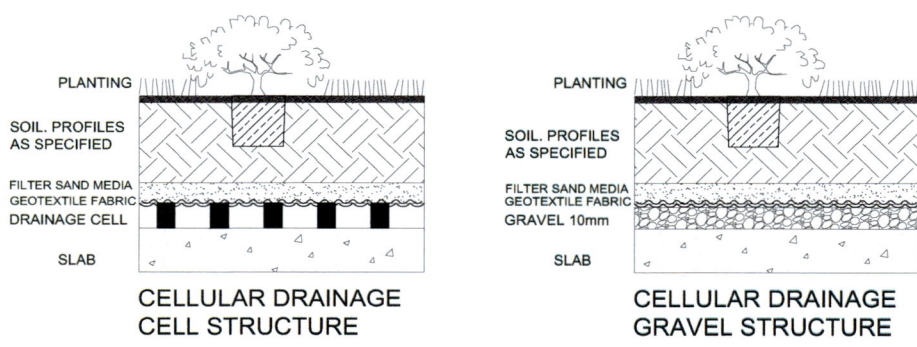

CELLULAR DRAINAGE
CELL STRUCTURE

CELLULAR DRAINAGE
GRAVEL STRUCTURE

(E) ## CELLULAR DRAINAGE
APPLICABLE FOR ON SLAB LOCATIONS NOT TO SCALE. DIAGRAMMATIC ONLY.

Fig. 4.7. Cellular drainage types. Options include (left) various manufactured 'cell' products, and (right) gravel structures.

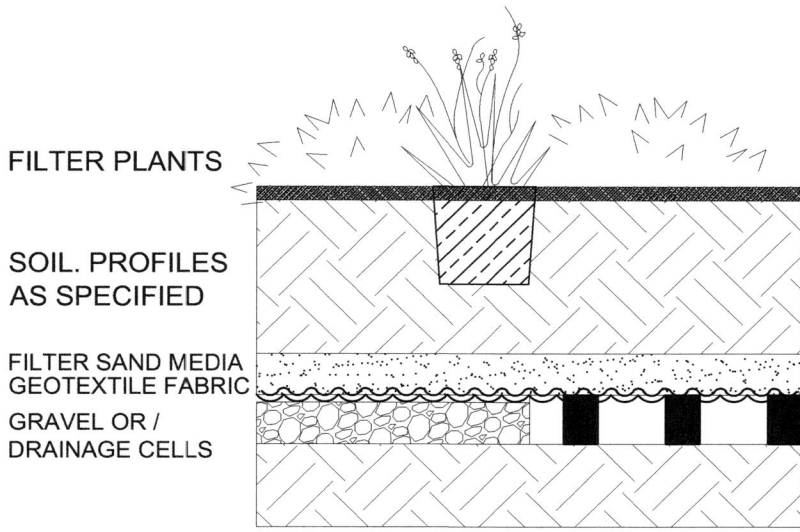

FILTER PLANTS

SOIL. PROFILES
AS SPECIFIED

FILTER SAND MEDIA
GEOTEXTILE FABRIC

GRAVEL OR /
DRAINAGE CELLS

FILTERING MEDIA
NOT TO SCALE. DIAGRAMMATIC ONLY.

Fig. 4.8. Drainage system designed as a filter.

4.11 SPECIFYING SOIL VOLUMES

Specification B4 includes a table on soil depths and a table for suggested soil volumes for trees. Volumes are generally specified for tree planting in urban situations where soil may be limited, and for medium to larger size plantings in confined growing conditions (such as raised planters or pots or where surrounded by hard-pave surfaces (i.e. in any place where there is no substantial 'break-out' zone for tree roots to utilise).

Local and other governing authorities often also state minimum tree soil volume requirements, which should be checked for compliance. Local authority minimum soil volume requirements vary quite substantially and may be a result of local climatic conditions or other cultural and political influences.

Eight factors that should be considered when determining soil volume requirements for trees in limited growing situations are shown in **Appendix C**, Fig. C1. See Fig. C3 for examples of the shared root system efficiencies that can be achieved.

All the eight key factors in Table 4.6 are linked and need to be suitably considered and balanced, meaning that all key factors should be addressed to ensure a tree, or trees, can be sustained for the designed expected lifespan.

Similarly, the Soil Volume Simulator is interlinked with these eight key factors (see **Appendix C**, Figs C1 and C4).

Refer to **Appendix C** and Fig. C4 for the Soil Volume Simulator approach method to determine parameters for trees in soil-limited spaces.

Table 4.6. The eight key factors in calculating soil volume requirement (refer also to Appendix C)

1. Available soil moisture and available water-holding capacity (and the depth to which that becomes limited)

2. Climate and climatic growing conditions

3. Physical soil properties and soil texture

4. Proposed soil ameliorations

5. Species selection

6. Shared root systems* and expected lifespan/replanting frequency

7. Maintenance, establishment and care, especially availability of irrigation

8. Available soil nutrients for plant growth including proposed soil ameliorations

* After Solfjeld (2009) and further studies; shared root systems can occur where trees of the same species or similar species can (but not always) fuse (or conjoin) feeder roots together or sometimes via mycorrhizal filament connections to enable shared soil resources (water and nutrients). Shared root systems can occur naturally in forest systems and can enable greater tolerance of a stand to stability, climatic and growing condition extremes (such as drought, wind or shallow soils), as well as provide more effective buffer for pathogen, fungus or bacterial imbalances. Refer to **Appendix C**, Fig. C3. Root system dieback during extreme drought/flood is also lessened. The dimensions of soil required for shared root systems to develop depend on adequate provision of 'the eight key factors' listed above and can be simulated with part 3 of the Soil Volume Simulator. A shared root system zone needs to be wide and deep enough to provide suitable growing conditions for tree roots to establish conjoined roots and trees to be spaced appropriately to promote conjoining of roots.

Appendix C, together with the freely available online Soil Volume Simulator, provides an intellectual process in the form of an online simulator to determine volumes specific to project constraints and parameters. A worked pdf of the Soil Volume Simulator can be easily generated and inserted into reports or presented to clients to demonstrate the method and design process of determining soil volumes for trees in limited spaces. Alternatively, the minimum and maximum range of soil volumes is tabulated in Figure C4.

4.12 SPECIFYING THE SOIL VOLUMES, PROFILE STRUCTURES AND DEPTHS FOR TREES IN AREAS WITH LIMITED SOIL

Specification B4 includes a template table that should be included in the soil specification (as applicable to your project) to outline the minimum approved volume, profile, depths and soil type.

Minimum soil volumes for trees are required only where soil volume is limited, or where there is risk of design encroachment or encroachment during construction of the minimum soil volume designed for.

The methodology for soil rooting volumes for trees in limited spaces, applying the eight key factors, and the Soil Volume Simulator can now be demonstrated and specified into the soil specifications. Refer to **Appendix C**.

4.13 CHOOSING THE PART III PRODUCT SPECIFICATIONS

After Part I and Part II of the specifications are selected, Part III, the product specifications, can be copied and pasted as appropriate to the landscape type that has been designed.

Guidance notes are provided throughout the performance specifications (Chapter 6) and preceding each of the Part III product specification types is a page with notes on how to use the example product specification.

4.13.1 Interactive component of specifications

Within the Part III product specifications themselves, an interactive (changeable) component exists for some of the specifications with regard to:

1. alkalinity and acidity pH levels in soils
2. phosphorus levels in soils (Table 4.7)
3. organic soil option.

The main reason for this 'interactive component' is to accommodate geographical locations with existing atypical pH and phosphorus levels that are outside the industry accepted 'optimum' range for landscape plant growth. The main consideration here is that any locally imported soil will also likely be out of the typical range for pH or phosphorus. The landscape designer must be aware of these soil environments and must design accordingly, and must also be responsible for deciding on using the local analysed soil product that may be outside the typical range in favour of transporting a non-locally sourced material (often at great environmental and monetary expense).

Note that no specific option has been given for acidic soils such as may be required for Ericaceous plants (azaleas, rhododendron and heather) and plants in the Theaceae family (such as camelias). Refer to Chapter 3, Table 3.3 for additional examples. Specifiers should note where such plants are used that the pH should be adjusted downward to around 5.0–5.8. Alkaline soil would preclude the use of such plant types. Such plants will also need lower P levels but not as low as the low P option. Specialist soils advice should be sought when changing the specification to suit specific plant types like this.

Important note: the specification examples provided in Chapter 6 cover only the common vegetation types used in landscape developments. Where deviation from these occurs or where specialist vegetation types are used, it is strongly recommended that the services of a soil technologist/scientist be employed. Specification of soil chemical and physical properties is an expert specialisation and easy to get wrong by the untrained. Also, when designing to the specifications using locally available materials, the landscape architect and their soil supplier are advised to use soil specialists, particularly for certification and validation.

Table 4.7.   An example of the interactive component of the specification where the landscape architect/designer must select the alkalinity and phosphorus component relevant to the project site and planting proposal

Note: this interactive table can be substituted in your specifications for other specific tolerances (such as salinity or acidity) depending on the typical soils in your local area.
Choose from the following alternatives based on the soil approach and design approach method (Section 4.7). Not all specifications have an interactive component.

Alkaline soils	Phosphorus-sensitive plants
☐ YES – Soils are alkaline or imported soils may be alkaline within the alkaline range from Table C2	☐ YES – Phosphorus-sensitive plants are included in the design. Phosphorus level must be in the low P range from Table C2
☐ NO – Soils are to be within the standard range of pH from Table C2	☐ NO – Phosphorus-tolerant plants have been chosen. Phosphorus levels must be in the standard range in Table C2

4.14 MULCHES

The recommended method of specifying mulches in Australia is to use *AS 4454 Composts, Soil Conditioners and Mulches*. However, the experienced landscape architect writing the specification should be aware that there are three different categories of mulch described in this standard being:

- particle size differentiation of mulch ('coarse' or 'fine')
- degree of decomposition of mulch
- phosphorus level in mulch.

4.14.1 Particle size differentiation

The two categories of mulch, as defined by particle size are:

1. *Coarse mulch:* equal to, or more than, 70% by mass in the shortest dimension to be retained by a 16 mm sieve.
2. *Fine mulch:* less than 20% by mass passing through a 5 mm sieve and less than 20% by mass in the shortest dimension to be retained by a 16 mm sieve.

The purpose of these particle size restrictions is to ensure that very fine compost material is not used as mulch. Such fine compost placed on the surface will actually restrict water entry into the profile and ensure that increased amounts of rainfall either run off directly or are absorbed by the mulch and evaporate again. Mulch must be coarse enough to freely admit water at high rates and then trap air to prevent the water evaporating.

4.14.2 Degree of decomposition

The second differentiation is the degree of decomposition or composting the mulch has been subjected to. These are summarised in Table 4.8.

Table 4.8.   The degree of decomposition of mulch

Stability category	Description	Examples	Issues
Raw mulch	Pure source product that does not need composting	Red gum chips, coloured woodchip, pine flake, pine bark and nuggets, sugarcane bagasse	Possible nitrogen draw-down. Usually highest quality appearance
Pasteurised mulch	Organic material subjected to a minimum of composting consistent with weed seed and disease eradication	Site chip mulch or line and road clearance chip*	Nitrogen draw-down, possible weed survival, lack of control of particle size (too many fines)
Composted mulch	Organic material composted for a minimum of 6 weeks with turning and screening	Composted council garden or 'green' waste. Construction and demolition timber	Very commonly available. Some litter contamination (plastics and metal). Dull appearance, nitrogen draw-down
Mature mulch	Organic material composted for a minimum of 12 weeks or until demonstrably stable with turning and screening	Composted council garden or 'green' waste. Construction and demolition timber	Not commonly available, costly to compost for such long periods

*Chipped vegetation from site and line clearance is often available and must be subjected to a pasteurisation as a minimum. Refer to Australian Standard AS 4454 for definitions.

The most commonly available mulches commercially will be raw mulches and composted mulches, both fine and coarse. In the authors' view, it is not necessary to specify mature mulch, but rather 'pasteurised and well-composted mulch'.

4.14.3 Phosphorus levels in mulch

A third differentiation is the amount of phosphorus in the mulch to prevent excessively high P levels being applied to P-sensitive plants (see Table 3.3). This does not apply to raw mulches because none of these will have excessive phosphorus. A pasteurised, composted or mature mulch recommended for P-sensitive plants must show less than 5 mg/L soluble phosphorus and less than 0.1% w/w total phosphorus.

4.15 SPECIFICATIONS FOR MULCH

4.15.1 Raw mulch

There are two parts to the specification of which both must be specified. These parts include:

1. the degree of fineness
2. the degree of composting, for example:

Raw compost: A pine bark mulch complying with the requirements for fine mulch and raw mulch from *Australian Standard AS 4454 Composts, Soil Conditioners and Mulches*.

   Note: landscape architects writing specifications in other countries will need to refer to their country's equivalent compliance standards.

4.15.2 Composted mulches

There are three parts to the specification for composted mulches being:

1. the degree of fineness
2. the degree of composting
3. the level of P, which must be specified if growing P-sensitive plants, for example:

Composted mulch: 'A recycled timber mulch complying with the requirements for fine mulch and composted mulch suitable for growing P-sensitive plants from Australian Standard AS 4454 Composts, soil conditioners and mulches' (Standards Australia 2012).

   Note: landscape architects writing specifications in another country will need to refer to their country's equivalent standards.

   The *recommended depth* that mulches should be applied is 75 mm, to be spread by hand after planting, and avoid contact of mulch with plant trunks to reduce likelihood of stem rot.

4.16 CREATING ARTIFICIAL AND SPECIALIST GROWING MEDIA

Artificial growing media usually refers to growing media that do not contain any soil, but in practice usually contain less than 20%. These growing media are specially developed for use in containers such as large pots, planter boxes, vertical walls and other confined root-zone applications. They are also different from soils physically in that they combine good water-holding capacity with high drainage and aeration. To achieve

Table 4.9.   Properties of commonly available materials for artificial growing media

Component	Description	Advantages	Disadvantages
Composted pine bark fines	Usually the less than 12 mm fraction of composted plantation pine	Lightweight, high water-holding capacity. By-product material	Low porosity, poor longevity (6 months max)
Composted coarse pine bark	Usually the 12–20 mm fraction of composted plantation pine	Lightweight, high porosity. By-product material	Low water-holding capacity, only moderate longevity (12–18 months)
Composted sawdusts	Composted sawdust from timber sawmills	Lightweight, high water-holding capacity. By-product material	Low porosity, poor longevity (6 months max), significant N draw-down
Coarse sand	A sand predominantly in the 0.5–2 mm size range	High density and weight, high porosity	High density low water-holding capacity, inert, low nutrient holding capacity. Mined material
Boiler ash, washed, screened or power station bottom ash	Coarse fused aluminosilicate material from high temperature combustion usually of coal	Lightweight, high longevity (years), reasonable water-holding capacity, low cost Recovered product	Not always regionally available, may contain fine silt
Perlite	Fused aerated silica balls	Very lightweight, high longevity (several years), very high water-holding capacity	Costly in comparison with ash
Pumice and scoria	Aerated volcanic extrusion	Very lightweight and high water-holding capacity, very high longevity (several years)	Only available regionally. Some scorias are quite dense
Diatomaceous earth	Fossilised siliceous skeletons of diatoms laid down in almost pure beds	Very lightweight, excellent water-holding capacity and longevity	Only available regionally, higher grades costly
Composted garden waste compost fines	Chipped council collection yard or garden waste composted and screened into compost fines and mulch fractions	Low cost, readily available, good nutrient content. Recycled material.	Can contain high coarse fraction if not screened
Coconut coir	The pith surrounding the shell of the coconut, milled into various sized fractions usually around 2 mm	Excellent water-holding capacity, very lightweight, reasonable longevity (2 years)	Moderately costly
Peat	Preserved remains of plants growing in saturated environments	Very low density and high water-holding capacity	Expensive, some very fine (poor aeration). Potential environmental damage at mining location
Vermiculite	Expanded mica clay	High cation and water-holding capacity	Cost, degrades reasonably quickly (12 months)
Plastic foams	Expanded balls of either polystyrene (Styrofoam™) or phenol formaldehydes	Very lightweight. Styrofoam™ is inert but formaldehyde foams hold water well. Increase porosity	Styrofoam is an inert filler and simply dilutes effective rooting volume

Component	Description	Advantages	Disadvantages
Zeolites	High cation exchange hard aluminosilicate clays	Good water-holding capacity and cation exchange	Variable quality, can be costly
Soil	Usually mined alluvium from river terraces rather than true topsoil	Provides mass and water-holding capacity, low cost	Can reduce porosity, heavy. A finite and becoming depleted resource.

these two mutually exclusive requirements, specialised, and often more costly, ingredients are used. The specifications in Part III E gives some suggested formulations based on experience with raw materials sourced on the eastern seaboard of Australia but, as with the rest of the specifications, these specifications are performance and product based and not recipe based. It is recommended that locally sourced and renewable or recycled materials are specified as a preference over those materials that are not.

Table 4.9 gives some background summary information on the properties of commonly available materials that will assist in understanding how the performance criteria may be met. One of the chief requirements in large installations is longevity, which is hard to specify, so some knowledge of longevity of ingredients is necessary.

Other materials may be locally available such as peanut hulls, coffee grounds, foundry sands and coke or slag from steel making. Each material needs to be assessed chemically and physically to see what contribution it may make to a mixture.

Seek expert advice from an experienced landscape technologist or local soil suppliers. Be wary of local soil suppliers: some are not as knowledgeable as they could be. Larger suppliers will often be more experienced because they will likely have supplied mixes for the intended applications before.

Guidance to using the soil performance specifications

5.1 USING THE TEMPLATES

The specifications aim to provide a tender-ready objective format that can simply be copied and pasted to specify soils for the great majority of landscape design projects.

The test requirements that must be met in the product specification tables should be achievable where a commercial soil supply industry and infrastructure of some reasonable sophistication is available. Where this is not available, variations and compromises have to be made (see Section 5.4 on variations). Also, some experienced professionals may argue with the target requirements, and indeed the test methods, we have set (see Section 5.5 on test methods).

This means that in some situations these 'copy and paste specifications' should be thought of more as pro formas or templates to give uniform structure and method to making soil specifications, and the target requirements may be varied, widened, simplified or added to as a result of local experience and the increase in knowledge that occurs over time.

5.2 STRUCTURE

Parts I and II of the specifications provide templates for the initial site investigation and preparation activities. They can be customised for such aspects as the size of the site (number of sampling sites to be used), the depth of the investigations (sites to be deeply cut may need deeper investigations than those indicated for surface soils) and the exact nature of the required subgrade specification.

Again, it is impossible to provide specifications precisely tailored to every possible situation a site or design presents, and the trained landscape architect or planner will be able to use them as a template to customise to the project, perhaps with the assistance of the site analysis team who will do the work.

Part III provides the supply specifications. These are in a three-part structure described in earlier chapters.

It is very important to note that Part III of the specifications is in a modular format that provides options that need to be selected to allow the use of *alkaline soil variants and low phosphorus variants*. This is where the interaction and influence on design is critical. If the site contains alkaline soil, we recommend selecting alkali- or lime-tolerant plants because it is often uneconomic to acidify alkaline soil. Some of the specifications allow for an organic soil option with significantly higher organic matter content. This is largely to keep these specifications consistent with AS 4419 that also describes an 'organic soil'.

Part A: 'Fit-for-purpose' performance description. This is in the form of a 'Fit-for-purpose' statement or general description of the intention that must be met. Where it may be argued that this conflicts with some of the product specifications, the performance specification should act as the guiding principle to vary the product specification.

Part B: Product specification (technical parameters). These are in the form of tables of requirements that must be met and are matched with the intention or performance specification.

Part C: Example components for the soil supplier. This is an informative section only and presents example formulations that may meet the compulsory performance and product specifications.

Where the soil is low in phosphorus, a design opportunity is given to use some of the spectacular array of South African and Australian plants in the Proteaceae family (or other plant species – refer to Table 3.3). In this instance, *select the 'low P' option in Part III* to make your specification very clear. Where the soil is high in phosphorus, such a treatment is ruled out because phosphorus cannot be removed from soil in the short term.

You *must select which option you want* to make it clear to the tenderers, contractors and soil supply companies.

5.3 SCOPE

Much consideration has gone in to ensuring that the 14 product specifications cover the majority of uses in landscape developments. The specifications listed should provide the majority of specifications required to assess site soils, prepare them for use and then specify soils to be constructed or altered for a given purpose. We have not provided separate specifications for improving site soils, it being implicit that they can be used in the end use performance and product specifications provided. Neither have we provided specifications for basic construction components such as drainage sand and gravel. Product specifications for mulches are well described in AS 4454 and do not need to be repeated here.

5.4 VARIATIONS AND NON-COMPLIANCE

Sometimes it is simply not going to be possible to meet the performance and product specifications or suggested components lists due to cost and availability constraints. An extreme example would be turf soils where high sand contents are required to provide the sort of performance ideal for the use. In some sites the soils are going to be clays and there is no budget for importing a lot of sand. Either modify the specification to allow for the use of site soil or stay with the 'ideal' specification and authorise or preferably get an expert in the field to authorise a variation with a clear explanation to the client of the possible consequences with a statement such as:

The use of clay soil on a sports field is likely to result in compaction and wear in heavily used areas such as around goal mouths and centre field.

In the case of minor variations, for example, the pH is 5.2 instead of 5.4, or the P level is 15 when 20 mg/kg is the specification, the variation may be authorised and the hold point lifted with or without a remedial action. It will usually take some experience and technical knowledge to confidently accept the responsibility for such variations. In some cases the specifications themselves will need to be varied as a result of specialist requirements or simple inability to meet the target ranges using available materials. Ideally an experienced horticulturist or soil scientist should be consulted as to just how far the ideal ranges can be varied without serious detrimental effect on the landscape.

5.5 TEST METHODS

The methods referred to in the specification tables are listed in Tables 6.33 and 6.34. These tables form part of Compliance **Specification G3** and are required to be inserted in the specifications as guidance for laboratories to use these or similar methods. If changes to the recommended methods are made upon expert advice then the tables need to be changed to reflect this.

There will no doubt be much argument and disagreement among the soil science community not just over the required ranges but also the test methods chosen. Soil scientists are not unlike economists or lawyers in this respect.

It is very important to take into account the following principles when mounting cases for and against any given method or required range:

1.   Most soil-testing methods are empirical and historically have been used to assess the fertiliser requirements of crop plants. Few or none have been calibrated for the growth of ornamental plants or for amenity landscapes.
2.   Experience with given test methods will vary between soil scientists. Unashamedly the tables of chemical and physical properties and the requirements given are best estimates from Simon Leake's nearly 40 years of soil analysis and consulting experience to the landscape industry. While there is some reference to established science in relevant texts and available standards they have not necessarily all been rigorously researched and nor can they be referenced to particular scientific papers. In this second edition, the authors have updated the physical and chemical properties based on industry feedback including soil suppliers, and developments in the industry in the last 10 years including level of quality of landscape outcomes and longevities, and updated testing methods.
3.   Local knowledge, available materials and planting types as well as test methods available locally will vary and the advice of local soil professionals should be sought in order to change or vary the technical product specifications. For example, there is no option for 'acid loving' plants which may do better at lower pH ranges than the ones we have suggested. If these are an important part of your planting list or local area, use the closest specification as a template and change the pH range (noting that the range has been altered, providing validation to that change and citing the template source).
4.   The authors are not precious about the ranges or test methods presented. It is more the method and format of the specifications that should be followed and the desire to introduce objective measurable criteria to specify landscape soils.

Local expert advice is available to assist in modifying or adding further prescriptive requirements. The Australian Society of Soil Science and, more widely, the International Union of Soil Sciences publish lists of soil scientist members and their associated expertise such as experience with urban soils. Equivalent societies and institutes for soil science exist in other countries for local expert advice and additional requirements.

5.5.1 The following test methods should be specifically noted

Mehlich 3 method

This is a fairly straightforward single acidic extract best suited to acidic soils and widely adopted in North America. Being strongly acidic it often gives higher numbers than other methods. It is not well suited to calcareous soils as it dissolves the lime giving falsely high calcium and phosphorus readings.

Calcareous soils

Alternative methods are given for the alkaline soil options that do not dissolve the lime.

It is very important that the whole specification including the test method recommendations are presented in tender documents.

That way all parties using and certifying to the specifications are properly informed, each section informing the parties in the following manner. Again note there are three essential parts to each specification:

1. *Performance specification (A):* this is mainly for the client and the landscape architect to understand and expect to get what they are asking for.
2. *Product specification (B):* this is an objective measurable method for the laboratories and competent experts to assess the product against the performance requirement and certify suitability.
3. *Informative section (C): this* provides help in an example and in plain English for the soil supplier or works contractor to get a head start on formulating the soil mix to meet the performance requirement on the product specification. It also provides advice to the laboratory on what test methods to employ and to the experts on how to deal with variations.

International users

Although written in Australia, making frequent references to 'P-sensitive Australian natives' as well as materials and components used in Australia, which may be largely non-applicable overseas, and using test methods possibly not available internationally, the fundamental method of making *performance specifications combined with technical product specifications* rather than recipe-based ones is relevant to all landscape developments.

Use the specifications as a template and use local knowledge of materials and testing methods to develop good objective specifications customised to local conditions. All we ask is a reference to this book.

5.6 HOW TO CONSTRUCT YOUR SOIL SPECIFICATION

Figure 5.1 summarises the components from Chapter 6 that must to be included in a landscape soil specification.

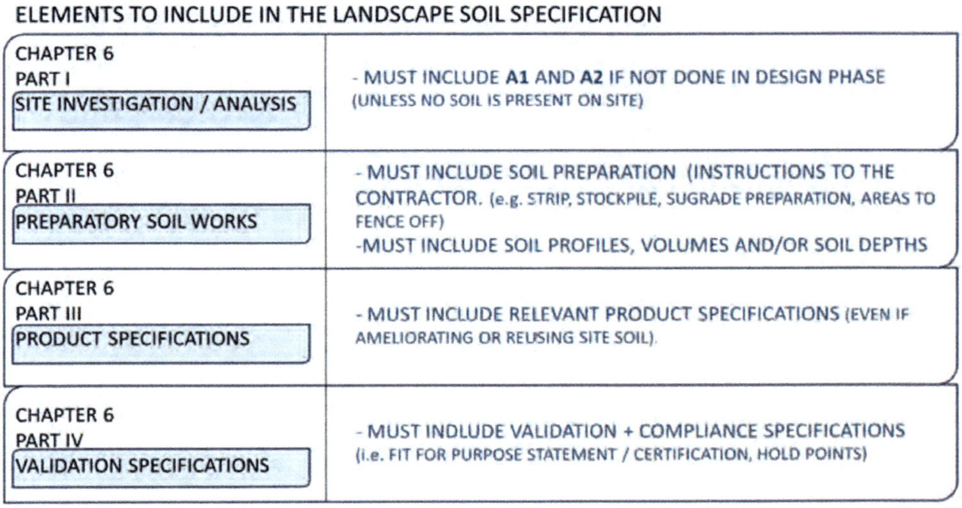

ELEMENTS TO INCLUDE IN THE LANDSCAPE SOIL SPECIFICATION

CHAPTER 6 PART I SITE INVESTIGATION / ANALYSIS	- MUST INCLUDE **A1** AND **A2** IF NOT DONE IN DESIGN PHASE (UNLESS NO SOIL IS PRESENT ON SITE)
CHAPTER 6 PART II PREPARATORY SOIL WORKS	- MUST INCLUDE SOIL PREPARATION (INSTRUCTIONS TO THE CONTRACTOR. (e.g. STRIP, STOCKPILE, SUGRADE PREPARATION, AREAS TO FENCE OFF) -MUST INCLUDE SOIL PROFILES, VOLUMES AND/OR SOIL DEPTHS
CHAPTER 6 PART III PRODUCT SPECIFICATIONS	- MUST INCLUDE RELEVANT PRODUCT SPECIFICATIONS (EVEN IF AMELIORATING OR REUSING SITE SOIL).
CHAPTER 6 PART IV VALIDATION SPECIFICATIONS	- MUST INDLUDE VALIDATION + COMPLIANCE SPECIFICATIONS (i.e. FIT FOR PURPOSE STATEMENT / CERTIFICATION, HOLD POINTS)

Fig. 5.1.   Elements to include in the landscape soil specification.

5.7 SMALL AND LOW BUDGET PROJECTS AND USING 'OFF THE SHELF' SOIL PRODUCTS: SIMPLE ADVICE AND BASIC TESTING

We recognise in the industry that there are some projects, clients or budgets that simply don't allow for full and proper soil survey and investigation. As outlined in this book, there are clear disadvantages and liabilities by not properly investigating soils. An evaluation of the risk of loss in money, time and resources for full or part landscape replacement, and lost client confidence or repeat business, might in some instances be worth taking.

Small project soil specifications such as courtyards and projects typically under 40 m^2 of garden size (lawn and garden beds) and generally on a residential scale usually only require a small volume of soil so testing can sometimes not be feasible. We still recommend the contract documents refer to the specifications in this book so at least there is some benchmark of quality that less scrupulous landscapers can be held to account with if problems occur due to the use of cheap poor quality soil or poor soil preparation.

Some soil laboratories are offering a basic and quick soil analysis, such as SESL Australia's 'Complete Soil Check' (Figs 5.2 and 5.3), specifically tailored to the home gardener and smaller landscape project.

At the time of writing, SESL Australia provides this basic analysis for approximately AUD$100 including postage. This particular soil analysis includes pH, salinity, texture, nitrogen, phosphorus, potassium, soil organic matter, micronutrients and soil cations exchange capacity. The results come with recommendations.

Typically, the small-scale landscape project is likely to implement the use of site soil with the combination of some imported soil. It can, however, be costly to order soil tests and generally an 'off the shelf' product is used. The key is to find a soil supplier that can provide the right soil for your situation.

Fig. 5.2. A basic soil-testing kit for the home gardener can assist with small and low-budget projects.

COMPLETE SOIL CHECK

Hi Elke, here's your CSC report for ▮▮▮▮▮▮▮▮▮▮▮▮▮ **stockpile**
Please present this report to your local Flower Power Garden Centre for product advice.

Analysis	Result		Requirement	Product
pH	6.3	🟩	Nil	
Salinity (dS/m)	0.07	🟩	Nil	
Nitrogen (mg/kg)	9.3	🟧	Fertiliser Recommended	Brunnings Complete Garden Fertiliser
Phosphorus (mg/kg)	230	🟥	Excessive - See fact sheet on CSC website	
Potassium (mg/kg)	210	🟩	Nil	
Organic Matter (%)	3.1	🟧	Compost Recommended	Supersoil Compost
Cation Exchange (cmol(+)/kg)	7.1	🟥	Compost Recommended	Supersoil Compost
Calcium (% eCEC)	77.5	🟩	Nil	
Magnesium (% eCEC)	13.9	🟩	Nil	
Sodium (% eCEC)	1.1	🟩	Nil	
Sulfur (mg/kg)	9	🟥	Sulfur Critical	Manutec Sulfur
Iron (mg/kg)	300	🟥	Excessive - See fact sheet on CSC website	
Manganese (mg/kg)	23	🟩	Nil	
Zinc (mg/kg)	4.9	🟥	Zinc Critical	Yates Citrus Cure Zinc & Maganese Chelate
Copper (mg/kg)	19	🟩	Nil	
Boron (mg/kg)	.3	🟥	Boron Critical	Searles Borax
Texture (class)	Sandy Loam	🟧	Potentially Droughty – Add compost or wetting agent	Supersoil Compost or Saturaid Wetting Agent

This soil report has been interpreted using the information and sample provided by the gardener. All products recommended have been laboratory tested for use by CSC. **We cannot guarantee the effectiveness of alternative products.** Any addition of compost will require an application of nitrogen to allow for sufficient nutrient to be supplied to plants. Please contact Flower Power for fertiliser recommendations and for advice if you have excessive elements in your soil.

🔴 **Product Critical**

🟠 **Product Recommended**

🟢 **Not Required**

This report was prepared for Elke Haege Thorvaldsen of Oxford St, Woollahra NSW 2025, Australia. Laboratory results are subject to natural variation. Requirements and products are based on information provided by the client and relate only to the plant type specified.

Fig. 5.3. A basic soil-test result can assist in providing initial basic useful, cost-saving information early on in a project and is generally sufficient information for small, low-budget projects.

5.7.1 How to specify for the small, low-budget projects

As the specifier (architect, landscape architect, landscape designer or home developer):

- work out the condition of the site soil preferably including some basic soil testing
- protect the site soil during construction such as by fencing off areas of the garden or by stockpiling the topsoil
- select plant species compatible with the existing soil conditions
- work out your finished raised levels and if you will also need soil to increase the finished levels of the garden. Generally allow for 250–300 mm topsoil (Horizon A) layer with a well-draining subsoil (Horizon B) beneath.

Again, we encourage the use of the soil specifications within this handbook (Chapter 6). Even if rigorous validation may not occur, they can be used to hold the landscaper to account if problems occur after establishment. By way of achieving rectification, tests can then occur to demonstrate compliance. If the soil is clearly not 'fit for purpose' this can be held as a defect to assist negotiation of a rectification.

Save costs by:

- stockpiling or fencing off parts of the garden during construction to protect the existing soil
- engaging a horticulturist or soil specialist to work out how you can ameliorate and condition your existing soil
- using the horticulturist or soil specialist to advise on improving or importing soil.

5.7.2 How to order soil for small, low-budget projects

Find a local and reputable soil supplier and discuss your soil requirements with the soil supplier. For example, describe:

- the intended purpose for the soil (e.g. exotic shrubs/hedges next to lawn)
- the existing condition of the soil (e.g. pH is 7.5, sandy and well draining, or it is being used as a construction site and the soil has been compacted or contaminated)
- the volume (area and depth) needed.

The soil supplier will recommend one of their products (all the better if they can state in writing that it is compliant or 'fit for purpose' with the soil specification type you require – see Chapter 6 for the list of typical landscape soil product types).

Request a soil-test sheet from the supplier and compare it to the product specifications in Chapter 6, noting that there may be some acceptable variance, or the product may be not acceptable.

Note: if you are only needing to condition/ameliorate your soil, then compare the example components (Part C, found in each example product specification in Chapter 6), factoring in your existing soil.

The proportion of the soil conditioner/amelioration is measured for the top 250–300 mm depth for topsoil (Horizon A) and could consist of:

- sandy loam soil
- clay loam
- composted soil conditioner conforming to your local standards (in Australia this is *Australian Standard AS4454 Composts soil conditioners and mulches*).
- lime or gypsum.

Note: without testing of the above mixture on site, variations and functionality can be varied, so its use is only feasible on a small project where plant and soil replacement or amelioration will not be too expensive if it needs to be redone including any plants that may have died.

5.7.3 What to ask the soil supplier

Many commercial producers of soil do not employ trained soil technologists to help with product design and quality. Some suppliers sell soils that will be detrimental or even kill the plants they are recommended for. Common problems we see are:

- lime and gypsum in recycled demolition soils, rendering them unsuitable for iron-inefficient plants
- too much quantity of rich manures, rendering them grossly excessive in nutrients to the point of salinity
- excessive use of compost. We see 'soils' with up to 70% compost in them. These go anaerobic and contain way too much phosphorus for native plants.
- high silt and clay content resulting in poor drainage, compaction and death through anaerobic conditions especially if also combined with too much organic matter.

Things to look for from a soil supplier are:

- How recent are the soil batch test results and are the results from the product that will be supplied to me? (Testing when the product is less than 3 months old is generally acceptable – after this, soil product properties can change.)
- Is the product suitable for my intended use/design/plant selections?
- What additional fertilisers, mulches or organic products are needed to meet the soil specification that you are after?

Don't buy soils from suppliers who cannot produce test certificates unless you are prepared to test them yourself before purchasing or installing. At least if the supplier provides a test result you have some come-back if problems occur.

5.7.4 Bare minimum properties to compare in the soil-test sheet from the soil supplier

As a minimum, compare the following to the relevant soil product specification in Chapter 6:

- pH
- permeability
- electrical conductivity (salinity)
- phosphorus (where phosphorus sensitive plants are being used, e.g. some Australian and South African native plants, and some northern European heathland species)
- texture (to estimate clay content)
- organic matter.

5.7.5 Variations and non-compliance to the specification

When comparing commercial soils with the intention of the specification chosen in Chapter 6, we cannot expect exact compliance. Target ranges are given in the specifications but minor non-compliances can still be considered 'fit for purpose'. Up to a 20% variation from the chemical soil tests (pH, electrical conductivity, phosphorus, organic matter) should be considered acceptable, and for physical tests (permeability) 30% variation should be considered compliant for small, low-budget landscapes.

The performance specifications

Part I Site investigation/analysis
- A1 Site soil investigation and characterisation
- A2 Site subgrade investigation and characterisation

Part II Preparatory soil works
- B1 Stripping and stockpiling
- B2 Site subgrade preparation
- B3 Imported subsoil
- B4 Soil schedules (profiles, volumes and depths)

Part III Product specifications
C Soils for turf and lawns
- C1 Passive amenity turf
- C2 Active high-traffic turf
- C3 Sports field turf

D Soils for gardens and mass planting
- D1 Mass planting soil
- D2 Garden bed planting soil
- D3 High fertility display and vegetable production topsoil
- D4 Advanced tree and vault soils

E On slab media
- E1 On slab soil media 'A' horizon
- E2 On slab soil media 'B' horizon
- E3 Low-density container and green roof
- E4 Ultra lightweight growing media 'A' only Horizon

F Specialist soils
- F1 Structural support soils
- F2 Raingardens biofiltration and stormwater soils
- F3 Wetland soils

Part IV Validation specifications
- G1 Quality assurance and control
- G2 Hold points and inspections
- G3 Compliance certification 'fit-for-purpose' statement and test method references

PART I A SITE INVESTIGATION/ANALYSIS

How to use the Part I specifications

Part I Site investigation/analysis comprises two typical specifications:

1. A1 Site soil investigation and characterisation (for largely intact soils)
2. A2 Site subgrade investigation and characterisation (for disturbed soils).

A1, or both A1 and A2, must be selected and included in your specification as a minimum requirement. Copy and paste the full specification (Parts A, B and C) to form part of your landscape specification document.

PDF templates of the soil specifications are available at <https://www.publish.csiro.au/book/8226>.

Note: these specifications are typical only, not site specific, and may require specialist advice where site conditions or performance criteria is outside the parameters of these typical specifications.

These typical specifications have been provided for use by proficient and experienced landscape design professionals who are able to determine when these specifications are suitable. Where there is any doubt, consult with appropriate experienced professionals, such as restoration ecologists, soil scientists, experienced landscape architects or consulting arborists.

Helpful tips for Part I

Existing site soil characterisation – decision pathway for Part I:

1. An intact or reasonably intact natural soil profile exists on the majority of the site and or well-established regrowth vegetation is present.
    _____ Go to Part I A1 Site soil investigation and characterisation.
2. The site is known to be or is obviously disturbed (e.g. landfill site, demolition, rehabilitation area with no or poor vegetation and no intact soil profiles).
    _____ Go to Part I A2 Site subgrade investigation and characterisation.

GENERAL BACKGROUND

Specification A1 provides a minimum standard for assessing the physical and chemical properties of an intact soil. The specification describes a method of assessing intact or substantially intact site soils for the purposes of adapting, stripping and reusing the soil for landscaping purposes. For small or low budget projects, Table B1 in **Appendix B** may be a suitable checklist for characterisation of the soil without the engagement of a soil scientist.

Specification A2 is used where a site is disturbed and unlikely to have intact profiles or useable topsoils. The investigation required as part of this specification will allow the experienced landscape technologist to determine whether the disturbed soil material present can be ameliorated for use as topsoil or subsoil or is so hostile it must be relegated as subgrade.

A checklist template for use during site visits can be found in **Appendix B**, Table B1, and **Appendix B3** Soil test request form. **Appendix B2** provides example soil analysis test results with recommendations to show an example of the information provided by the soil scientist following analysis.

Certification and 'fit-for-purpose' statement to be submitted (with any soil amendments needed). Refer to **Specifications G1** and **G2** for the testing and certification requirements.

The following is a typical specification that contains the elements required to ensure adequate investigation of the site soil before finalising design works. It can be used as a format for preparing or tailoring a site soil investigation specification for any project.

SPECIFICATION A1: SITE SOIL INVESTIGATION AND CHARACTERISATION

Specification A1 is for site analysis for substantially intact sites (the topsoil and the subsoil are present). Refer to **Appendix A** for a guide on taking *in situ* soil samples and the number of soil samples.

Site soil survey

A survey of the site's soil resource must be conducted. This usually requires the services of a trained soil scientist. The following information should be gathered as a minimum.

The uniformity or otherwise of the residual surface materials must be determined to 500 mm minimum depth. For larger tree stock, it is recommended to assess the profile to 1 m depth.

1. The morphology (texture, structure and colour) of the main types of surface materials present and their horizon designations should be determined.
2. The depths of each soil horizon to rock or parent material should be measured if possible.
3. Any physical limitations posed by the materials (stoniness, clay, poor drainage) should be assessed.
4. Samples representative of the main types of surface horizon (topsoil) material present must be analysed for the following properties, as a minimum:
   a. pH
   b. salinity
   c. cation exchange properties, including Ca:Mg ratio
   d. plant available nutrient contents: P, N, S, Fe, Mn, Zn, Cu, B
   e. dispersibility and aggregate stability
   f. organic matter
   g. texture or particle size analysis
   h. stone content.
5. Samples representative of the main types of subsurface horizons (subsoil) present must be analysed for the following properties, as a minimum:
   a. pH
   b. salinity
   c. cation exchange properties
   d. dispersibility and aggregate stability
   e. texture or particle size analysis
   f. stone content.
6. Where there is any suspicion of salinity, a deep subsoil sample (to around 800 mm depth) must be taken and also analysed for subsoil properties as above.
7. The consultant must provide a report identifying as a minimum:
   a. a description of the field condition of the surface materials soil (results of the field survey)
   b. interpretation of test results
   c. a statement of 'fitness for purpose' as topsoil, subsoil or subgrade
   d. recommendations for reuse, amelioration or improvement of both topsoil and subsoil.
8. The report must include comments and recommendations on the following details:
   a. the depth of each soil horizon

b. the morphology (texture, structure and colour) of at least the A and B horizons
c. the presence of any inclusions (ironstone, manganese pellets, lime concretions, visual contaminants such as glass, plastics or bricks)
d. the soil type or classification of the soil(s) present
e. any areas of disturbed, filled or altered conditions that render the soil unusable or raises special requirements
f. the depth of the topsoil and any variation in depth for stripping purposes
g. recommended topsoil stripping depths and stockpiling methods
h. any limitations imposed by the chemical and physical properties of the soils
i. the means by which the soils may be ameliorated or improved for various landscape purposes.

SPECIFICATION A2: SITE SUBGRADE INVESTIGATION AND CHARACTERISATION FOR DISTURBED SITES

Specification A2 is for site analysis for substantially disturbed sites (the topsoil and subsoil are disturbed or not present). Refer to Section 3.6 for soil contamination, and **Appendix A** for a guide on taking *in situ* soil samples and the number of soil samples.

Site soil survey

A survey of the site's soil resource must be conducted. This usually requires the services of a trained soil scientist. The following information should be gathered as a minimum:

1. the uniformity or otherwise of the residual surface materials must be determined to 500 mm minimum depth. For larger tree stock, it is recommended to assess the profile to 1 m depth.
2. the morphology (texture, structure and colour) of the main types of surface materials present and their horizon designations should be determined, if applicable
3. assessment of any physical limitations posed by the materials (stoniness, clay, poor drainage)
4. analysis of samples representative of the main types of surface materials present for the following properties as a minimum:
   a. pH
   b. salinity
   c. cation exchange properties
   d. plant available nutrient contents: P, N, S, Fe, Mn, Zn, Cu, B
   e. dispersibility and aggregate stability
   f. organic matter
   g. texture or particle size analysis
   h. stone content.
5. Samples representative of the main types of subsurface horizons (subsoil) material present must be analysed for the following properties, as a minimum:
   a. pH
   b. salinity
   c. cation exchange properties, including Ca:Mg ratio
   d. dispersibility and aggregate stability
   e. texture or particle size analysis
   f. stone content.
6. Where there is any suspicion of salinity, a deep subsoil sample (to around 800 mm depth) must be taken and also analysed for subsoil properties as above.
7. Where there is suspicion of contamination or plant toxicities include testing for heavy metals and other contaminants. (Refer to Sections 3.3-3.7.)
8. The consultant must provide a report identifying as a minimum:
   a. a description of the field condition of the surface materials soil (results of the field survey
   b. interpretation of test result
   c. a statement of 'fitness for purpose' as topsoil, subsoil or subgrade
   d. recommendations for reuse, amelioration, improvement or burial as subgrade.
9. The report must include comments and recommendations on the following details:

a. the morphology (texture, structure and colour) of the main surface materials present
b. the presence of any inclusions (ironstone, manganese pellets, lime concretions, visual contaminants such as glass, plastics or bricks)
c. all areas of disturbed, filled or altered conditions
d. any limitations imposed by the chemical and physical properties of the soils
e. the means by which the soils may be ameliorated or improved for use at topsoil or subsoil
f. what, if any, soil materials must be imported for the achievement of landscape aims.

PART II B PREPARATORY SOIL WORKS

How to use Part II

Part I must be carried out in order to correctly specify components in Part II. Part II provides three preparatory options being:

1. B1 Stripping and stockpiling (for soil that can be recovered)
2. B2 Site subgrade preparation (for subgrade that can be prepared as a subsoil)
3. B3 Imported subsoil (for instances where subsoil needs to be imported).

A fourth specification (B4) must be included in all soil specification documents:

4. B4 Soil schedules (profiles, volumes and depths).

Note: B4 requires the specifier to choose, select and fill in tables as appropriate (it is not a simple copy/paste specification). It may require the services of a trained and experienced urban soil scientist.

Specifications B1, B2 and **B3** may or may not be required for your specification. Some projects may require all (B1, B2, B3 and B4) specifications. Copy and paste the entire specification into your landscape specification document as appropriate.

PDF templates of the soil specifications are available at <https://www.publish.csiro.au/book/8226>.

Note: these specifications are typical only, not site specific, and may require specialist advice where site conditions or performance criteria is outside the parameters of these typical specifications.

These typical specifications have been provided for use by proficient and experienced landscape design professionals who are able to determine when these specifications are suitable. Where there is any doubt, consult with appropriate experienced professionals such as restoration ecologists, soil scientists or experienced landscape architects/consulting arborists. The following is a typical stripping and stockpiling specification that contains the elements required to ensure adequate measures are met in order for stripping and stockpiling to be effective, useful and successful. This specification can be used as a template for any project as appropriate.

Helpful tips for Part II

Preparatory soil works: decision pathway for Part II

1. The current soil landscape *can be recovered* to form functional soil profiles.
    _____ Go to B1 Stripping and stockpiling.
2. The current subgrade *cannot* be ameliorated to form topsoil but can be prepared as subsoil *before placing imported topsoil.*
    _____ Go to B2 Site subgrade preparation.
3. The current subgrade *cannot* be ameliorated (to form site subsoil soil) or the levels need to be raised *before placing imported topsoil* to meet the finished levels.
    _____ Go to B3 Imported subsoil.

SPECIFICATION B1: STRIPPING AND STOCKPILING

General background

Prior to stripping, a soil investigation should take place to clearly define varying soil typologies, topsoil and subsoil depths, and inform the stripping and stockpiling process.

Stripping and bulk earthworks should only occur when soil is at field capacity to prevent structural damage. Post rainfall, allow 48 hours for soil to drain and regain strength to minimise the risk of structural damage of the soil. Ensure a plan is followed to minimise multiple handling of the soil.

Stripping and stockpiling of topsoil should occur immediately before bulk earthworks and be done in such a manner as to minimise erosion and sediment loss from site. Preparation is necessary to ensure that rubbish and foreign matter is minimised in the stripped soil. Stockpiles must be located in a convenient place away from any risk of running water and subject to suitable erosion control measures. They must be protected from contamination during the construction process and records kept of their location and type of soil, if any, they contain.

Stripping

Preparation:	• Following necessary approvals, clear all debris, including demolition waste, timber, rubbish wire fences, rock and gravelled driveways.
	• Clear trees and shrub growth and slash if necessary.
	• Clear pasture and weed growth. For small sites, where possible remove weeds by hand tools, flame torch or by other mechanical means, and only as a last priority and not within waterways or environmentally sensitive areas. Wet-spray (non-aerosolised) with a broad spectrum herbicide at manufacturer's rate, allowing 1–2 weeks to obtain kill before stripping. Obtain written approval from site superintendent and authority (as suitable) of proposed clearing method prior to works.
Stripping:	• Avoid the inclusion of subsoil in topsoil stripping, adjust depth accordingly.
	• Strip topsoil to < *Insert mm* > depth or
	• Strip topsoil according to recommendations of < *Insert report name* >.

Stockpiling

Stockpile construction and management:	• Locate stockpiles 5 m or more from concentrated water flows (including drainage lines, roadways).
	• Locations should have less than 10% slope.
	• Locate greater than 8 m from any retained trees.
	• Protect upslope using diversion drains.
	• Protect downslope sediment loss using sediment control structures (silt fencing or other approved method).
	• Stockpiles must be no higher than 2 m but may be flat topped.
	• Label stockpiles with origin and date.
	• Protect stockpiles from waste and rubbish dumping and encroachment of works.
	• If stockpiles are to be in place longer than 3 months, sow with a seasonally appropriate annual cover crop.

SPECIFICATION B2: SITE SUBGRADE PREPARATION

The following is a typical site subgrade specification that contains the elements required to ensure adequate preparation of the subgrade before topsoil placement. It can be used as a format for preparing a subgrade treatment specification for any project.

Before laying topsoil, the following subgrade treatment must be applied to all finished subgrade areas:

1. Ensure levels are reduced with adequate falls and drainage ready to receive the required overall soil depths (see **Specification B4**).
2. Remove rocks > 100 mm diameter.
3. Remove rubbish such as construction generated waste, plastics, metals and glass.
4. Apply gypsum and lime according to the following schedule of gypsum and lime requirements to ameliorate the subgrade. Insert the application rate from the soil reports (**Specifications A1** and **A2**) or obtain further soil tests. Do not apply gypsum or lime without prior approval from an agronomist or soil scientist. Incorrect application can cause undesirable pH levels and/or salinity.

Landscape treatment	Gypsum (g/m^2)	Agricultural lime (g/m^2)
General grassing, native mass planting (grasses, shrubs and trees)	< *Insert rate* >	< *Insert rate* >
High-quality amenity turf, housing lots, display beds	< *Insert rate* >	< *Insert rate* >

5. Chisel, disc plough or use an excavator with a tine attachment to loosen the subgrade and mix the ameliorants to 200 mm depth to incorporate.
6. Harrow to break up clods but do not smooth (leave the surface 'keyed' to accept the rebuilt soil profile).

Note: use an excavator with a tine or ripping blade for operations on the steeper batters or where access is difficult.

Refer to soil reports for any other required actions (e.g. bucket screening to 50 mm).

SPECIFICATION B3: IMPORTED SUBSOIL

This specification is to be used to manage the importation of subsoil, or its manufacture from on-site materials where present subgrades do not provide sufficient quality to qualify as a rooting medium to provide rooting depth sufficient for larger plantings.

Part A. 'Fit-for-purpose' performance description

Generally a low organic matter soil that is well balanced chemically, is not saline or sodic or excessively acidic or calcium deficient, and not dispersive. It is designed to provide improved rooting depth for larger plantings and reduce the likelihood of water-logging. It may be made up using site subsoil or fill materials or a blend of both. It is not generally considered to require the application of fertiliser to subsoil but must be low in P if used for P-sensitive plantings.

Choose from the following alternatives based on the soil approach and design approach method:

Alkaline soils	Phosphorus-sensitive plants
☐ YES – Soils are alkaline or imported soils may be alkaline within the given alkaline range (Refer to Table 6.2) ☐ NO – Soils are to be within the standard range of pH from Table 6.2	☐ YES – Phosphorus-sensitive plants are included in the design. Phosphorus level must be in the low P range from Table 6.2 ☐ NO – Phosphorus-tolerant plants have been chosen. Phosphorus levels must be within the standard range from Table 6.2

Note: if the above selections are not chosen, the landscape contractor/soil supplier must communicate with the landscape architect/specifier for determination.

Part B. Product specification (technical parameters)

Generally the soil must be free of 'unwanted material' and must meet all the requirements of Tables 6.1 and 6.2. Where variations from these requirements occur, refer to **Specifications G1, G2** and **G3**.

> Certification and 'fit-for-purpose' statement to be submitted (with any soil amendments needed). Refer to **Specifications G1, G2** and **G3** for the testing, 'fit-for-purpose', certification and test method requirements.

Table 6.1.   Physical properties

Property	Units	Target range
Texture, preferred range	n/a	Sandy loam to clay loam
Permeability (Mc&J) or if texture is heavier than fine sandy clay loam, use 'estimated permeability'	mm/h	> 20. At 16 drop compaction level
Emerson aggregate class	n/a	> 4
Large particles (method ref. AS7755 5.4) in the largest dimension		
2–10 mm	% w/w	< 20
10–20 mm	% w/w	< 10
> 20	% w/w	< 10
> 50 mm	% w/w	< 2
Visible contaminants > 2 mm*	% w/w	0–0.5

*Of which plastics: 0–0.25, of which artificial sharps: 0 in 1.0 kg of air-dried soil.

Table 6.2. Chemical properties

Property	Units	Target range
Water repellence	seconds	60 s (water)
pH in water (1:5) standard range	pH units	5.4–6.8
pH in CaCl$_2$ (1:5) standard range	pH units	5.2–6.5
pH in water (1:5) alkaline range	pH units	6.8–8.0
pH in CaCl$_2$ (1:5) alkaline range	pH units	6.5–7.5
Electrical conductivity (1:5)	dS/m	< 0.5
Chloride	mg/kg	< 200
Phosphorus – P-tolerant or standard plants acid soils method	mg/kg	< 50
Phosphorus – P-sensitive plants alkaline soils method 9B1 or 9C1	mg/kg	< 20
Exchangeable sodium (Na)	% of ECEC	< 15
Exchangeable potassium (K)	% of ECEC	3–10
Exchangeable calcium (Ca) method 18F1 or 15A1 in alkaline soils	% of ECEC	Normal soil 60–80 Alkaline soil 70–90
Exchangeable magnesium (Mg)	% of ECEC	15–25
Ca:Mg ratio	Ratio	1.5–8

Part C. Example components for the soil supplier

The following table outlines suggested components that may meet the physical requirements of this specification. This is not part of the product specification. It is an example for the edification of the soil supplier of what might meet the product specification.

Example components (likely to meet the physical requirements of this specification):

Sandy or sandy loam soil	20–40% v/v
On-site clay loam or clay subsoil	30–60% v/v

Base level requirements for fertilisers (to be verified by laboratory testing and per agronomist's report):

Lime and/or dolomite or	2 kg/m^3
Gypsum	2 kg/m^3

For the purposes of tendering, the contractor must allow for the inclusion of the above soil amendments, but the specific amendments required must be verified by laboratory testing and agronomist recommendations.

See **Specifications G1**, **G2** and **G3** for validation, certification and test method specifications.

SPECIFICATION B4: SOIL SCHEDULES (PROFILES, VOLUMES AND DEPTHS)

How to use the B4 soil schedules

The following is a set of templates that may be used as a format and filled in accordingly for any project. **Specification B4** must form part of the landscape specification and may include some or all of the parts (as deemed appropriate by experienced industry-qualified professionals).

The B4 soils schedules (profiles, volumes and depths) include:

- profile horizon
- schedule of soil depths
- template for tree volumes and depths.

Note: these specifications are typical only, not site specific, and may require specialist advice where site conditions or performance criteria is outside the parameters of these typical specifications.

These typical specifications have been provided for use by proficient and experienced landscape design professionals who are able to determine when these specifications are suitable. Where there is any doubt, consult appropriate experienced professionals such as restoration ecologists, soil scientists or experienced landscape architects/consulting arborists.

> **Helpful tips for Part II, Specification B4**
> Next step: State the horizon structures and soil depths needed for the project and volume requirements (e.g. tree pit volumes).
> _____ Continue with **Specification B4** soil schedules (profiles, volumes and depths).

Specification B4 soil schedules (profiles, volumes and depths)

Choose from the table below which profile horizon structure each soil type must comprise A/C or A/B/C horizons.

To determine which profile is suitable to choose, refer to Section 4.8. Different landscape types, design objectives and site conditions will determine which horizon is suitable, described in Sections 4.2–4.7.

Profile horizon structure

The following landscape types must have the following profile horizon structure:

Landscape type	Profile horizon structure
Turf	*<Add either A/C or A/B/C horizons>*
Shrubs/mass planting Trees	*<Add either A/C or A/B/C horizons>*
<Add more landscape types as required for your project and outline the profile horizon structure>	A/B/C horizons

SCHEDULE OF SOIL HORIZON DEPTHS

The following table provides recommended schedules for horizon arrangements and depths for different landscape types. Use these as a base and alter or provide exact specifications in a table as in the worked example that follows.

Constructed soils must comprise of a horizontal layered arrangement (as seen in natural soil profiles). Some landscapes will require two horizons, and others three horizons. This table outlines the profile arrangements that must be incorporated into the project.

Landscape type	Minimum total soil depth	Minimum recommended Topsoil depth (A horizon)	Typical minimum subsoil depth or depth of ameliorated subgrade (B horizon)
Turf	200 mm	150 mm	0–150 mm
Shrubs/mass planting	400 mm	250 mm	0–200 mm
Trees over 45 L or trees with shrubs	450 mm	250 mm	0–200 mm
Turf with trees	Varies: see A and B horizons	150 mm to taper down to 300 mm around tree	200 mm below root ball
Horticultural display	450 mm	300 mm	150 mm
Trees in limited soil conditions, e.g. streets, verges, tree pits, rooftops	400–600 mm	Refer to tree volume tables (refer to **Appendix C5**, Table C4)	200 mm below root ball

Notes: depths given are to be accepted as minimum depths and may be exceeded. Topsoils must not be increased to anything more than 400 mm.

Where specified, mulch depth must be 75 mm for all landscape types except with turf (which must comprise topdressing as outlined in the turf specifications C1, C2 and C3.

For clarification of locations of each landscape profile type, refer to the landscape plans and contact the landscape architect. Multiple landscape types and profile horizon structures may be required.

For tree planting where soil rooting volumes are limited, refer to the specification table 'Tree rooting volumes' on p. 115.

Soil profile horizons example

The following is an example schedule of horizon profile structure and depths.

A housing estate is to be landscaped with individual lots grassed with a border garden at the front, road verges planted to trees and shrubs and display beds at axis points such as roundabouts. The following soil depth schedule will be included in the tender documents.

Soil profile horizons specification

The following table outlines the landscape types included in the landscape plan and the depths of each horizon that must be installed in accordance with this specification.

Planting area	Total soil depth (mm)	Topsoil depth (mm)	Subsoil depth (mm)
1. Turfed lots	150	150	0
2. Lot border gardens	450	250	200
3. Verges mass planting and trees	400	200	200
4. Display beds	500	300	200

TREE ROOTING VOLUMES TABLE

The following table may be used as a template to outline the tree soil rooting volumes required. This is often needed for tree planting, trees in verges, vaulted tree pits, rooftop applications and other urban situations where soil volumes and depths are limited.

To determine the tree rooting volume requirements and consider the eight key determining factors, refer to **Appendix C1**, Fig. C1 and **Appendix C2**, Table C2.

The following table is an example of a list of typical minimum soil volumes for tree planting types on a site.

Tree type and location in project and approved installation pot size	Minimum soil volume per tree (m³)	Topsoil depth (A)/ subsoil depth (B)/ subgrade depth (C)	Notes: e.g. structural soils, drainage + irrigation requirements, maximum depth allowed
Street trees (e.g. 100 L)	Add min. volume, e.g. 30 m³	A: add e.g. 500– 560 mm B: add e.g. 300 mm C: add e.g. 200 mm	Add notes, e.g. depth between 400–600 mm, subsoil drainage as specified.
Trees in carpark (e.g. 75 L)	Add min. volume, e.g. 27 m³, for individual tree/ planting zone	A: add (mm) B: add C: add	Add notes, e.g. connected soil trench in median catchment swale as on plans.
Trees and screen hedges on rooftop (e.g. 45 L)	Add min. volume, e.g. 18 m³ (less/ tree for shared rooting zones/ hedges)	A: add e.g. 350 mm B: add e.g. 300 mm C: add e.g. 200 mm	Excludes depths of drainage, drainage cell and waterproof membranes. Trees may require anchoring and stabilising.
Containerised trees/ trees in planters/ rooftop trees (e.g. 75 L)	Add min. volume, e.g. 20 m³ (less/ tree for shared rooting zones/ hedges)	A: add e.g. 350 mm B: add e.g. 300 mm C: add e.g. 200 mm	e.g. mounded, tree anchors, lightweight soil requirements (refer to **Specification E3**).
Large feature trees in structural soils or cells in plaza (e.g. 200 L)*	Add min. volume, e.g. 35 m³	A: add e.g. 650 mm B: add e.g. 200 mm C: add e.g. 200 mm	Irrigate. Note: provide A/B/C soil profile in tree planting hole. Structural cells to be located only where hardpave or risk of compaction may occur.

Note: refer to **Appendix C:** Soil rooting volumes for trees to inform and adapt this table to suit the eight influencing factors in determining minimum soil rooting volumes. A reduced soil volume may be possible by using the Soil Volume Simulator to inform this process.

Note: refer to the local guidelines and regulations for minimum volumes and specifications.

*When using structural soils, the volume of rock or structural plastic cells (e.g. Strata Vault © or other) must be subtracted to determine soil rooting volume. However, when using structural cells, the topsoil depth can often be increased due to increased air at depth, and hence promote deeper tree root growth, provided there is adequate drainage and irrigation.

PART III C SOILS FOR TURF AND LAWNS

How to use the Part III C specifications

There are three typical specification options for turf and lawns. These are:

- C1 Passive amenity turf (for moderate levels of pedestrian traffic)
- C2 Active high-traffic turf (for general sports playing fields)
- C3 Sports field turf (for high-performance sports).

If your design calls for turf or lawn areas, select the typical turf soil specification as appropriate to your design intent and add all the parts of that specification (Parts A, B and C) to form part of your landscape specification document.

PDF templates of the soil specifications are available at <https://www.publish.csiro.au/book/8226>.

Note: these specifications are typical only, not site specific, and may require specialist advice where site conditions or performance criteria is outside the parameters of these typical specifications.

These typical specifications have been provided for use by proficient and experienced landscape design professionals who are able to determine when these specifications are suitable. Where there is any doubt, consult with appropriate experienced professionals such as soil scientists or experienced landscape architects/consulting arborists.

The actual permeability to choose for sports fields depends on rainfall intensity of the area. 50 mm/h may be acceptable for sports fields in Melbourne with more gentle, usually winter, rainfall but 200 mm/h would be chosen for climates like Cairns with intense tropical storms. Major sports field design and construction is a highly expert process and it is recommended the landscape architect outsource this to experienced sports field soil professionals.

SPECIFICATION C1: PASSIVE AMENITY TURF

Part A. 'Fit-for-purpose' performance description

Generally, this requires a sandy loam 'turf underlay' topsoil mix designed to provide moderate resistance to compaction in public and other amenity turf areas subject moderate levels of pedestrian traffic. The specification is not suitable for active recreational areas and is not generally considered suitable for construction of playing fields, even with specific turf management practices to prevent compaction. The blend provides superior water-holding capacity to **Specification C2** soils.

Part B. Product specification (technical parameters)

Generally the soil must be free of 'unwanted material' and must meet all the requirements of Tables 6.3 and 6.4. Where variations from these requirements occur refer to **Specifications G1** and **G2**.

> Certification and 'fit-for-purpose' statement to be submitted (with any soil amendments needed). Refer to **Specifications G1** and **G2** for the testing and certification requirements.

Table 6.3.   Physical properties particle size analysis

Property	Units	Target range
Particle size distribution		
2.0 mm (fine gravel)	% retained by mass	< 10
1.0 mm (very coarse sand)	% retained by mass	< 10
0.5 mm (coarse sand)	% retained by mass	10–30
0.25 mm (medium sand)	% retained by mass	20–40
0.1 mm (fine sand)	% retained by mass	10–30
0.05 (very fine sand)	% retained by mass	5–15 (max 25% combined very fine sand, Si + Cl)
0.002 mm (silt)	% retained by mass	< 8 (Si + clay combined) 8–15
< 0.002 mm (clay)	% retained by mass	3–8
Large particles	% by mass	2–20 mm = < 10%, > 20 mm = 0%
Organic matter content	% w/w	2 to 6
Permeability (ASTM)	mm/h	32. At 16 drop compaction level
Water repellence	seconds	< 60

Table 6.4.   Chemical properties

Property	Units	Target range
pH in water (1:5)	pH units	5.4–8.0
pH in CaCl$_2$ (1:5)	pH units	5.2–7.5
Electrical conductivity (1:5)	dS/m	< 0.5
Exchangeable Na percentage	% of eCEC	< 15
Exchangeable Ca:Mg ratio	Ratio	3–9
Available phosphorus	mg/kg	
Acid soils method		50–150
Alkaline soils method		20–50
Available nitrogen (N as nitrate)	mg/kg	20–60

Part C. Example components for the soil supplier

The following table outlines suggested components that may likely meet the physical requirements of this specification. This is not part of the product specification. It is an example for the edification of the soil supplier of what might meet the product specification.

Example components (likely to meet the physical requirements of this specification):

Medium–coarse grade washed sand. Sandy loam soil or site soil	30–50% by volume	E.g. 5 parts washed
Composted soil conditioner conforming with	40–60% by volume	sand/4 parts site
AS 4454	10% by volume	soil loam/1 part
		AS 4454 compost

Base level requirements for fertilisers (to be verified by laboratory testing and per agronomist's report):

Lime and/or dolomite	2 kg/m^3 at mixing
Balanced compound NPK turf starter fertiliser Minor	0.5 kg/m^3 or 50 g/m^2 after placement
and trace elements	300 g/m^3 at mixing

For the purposes of tendering, the contractor must allow for the inclusion of the above soil amendments, but the specific amendments required must be verified by laboratory testing and agronomist recommendations.

See **Specifications G1**, **G2** and **G3** for validation, certification and test method specifications.

SPECIFICATION C2: ACTIVE HIGH-TRAFFIC TURF

Part A. 'Fit-for-purpose' performance description

Generally, a sandy, well-drained 'turf underlay' topsoil mix designed to provide resistance to compaction, rapid drainage but with adequate water-holding capacity to sustain turf growth. The specification is not as rigorous as a full USGA premium-grade playing field specification and is intended for moderate to high levels of use in competitive sports. The narrow fines specifications is considered important in meeting the shear strength requirement without risking undue compaction, but precedence will be given to meeting the shear strength and permeability test requirements.

Part B. Product specification (technical parameters)

Generally, the soil must be free of 'unwanted material' and must meet all the requirements of Tables 6.5 and 6.6. Where variations from these requirements occur refer to **Specifications G1** and **G2**.

> Certification and 'fit-for-purpose' statement to be submitted (with any soil amendments needed). Refer to **Specifications G1** and **G2** for the testing and certification requirements.

Table 6.5. Physical properties particle size analysis

Property	Units	Target range
Particle size analysis		
2.0 mm (fine gravel)	% retained by mass	< 3
1.0 mm (very coarse sand)	% retained by mass	< 10
0.5 mm (coarse sand)	% retained by mass	10–30
0.25 mm (medium sand)	% retained by mass	20–40
0.1 mm (fine sand)	% retained by mass	20–30
0.05 (very fine sand)	% retained by mass	< 15 (max 20% combined vfs, Si + Cl)
0.002 mm (silt)	% retained by mass	< 8 (Si + clay combined 5–8%)
< 0.002 mm (clay)	% retained by mass	2–6
Large particles	% by mass	2–10 mm < 2%, > 10 mm 0%
Organic matter content	% w/w	2–5
Permeability*	mm/h	50–120
Water repellence	seconds	< 60

*ASTM F1815–1997

Table 6.6. Chemical properties

Property	Units	Target range
pH in water (1:5)	pH units	5.4–8.0
pH in CaCl$_2$ (1:5)	pH units	5.2–7.5
Electrical conductivity (1:5)	dS/m	< 0.5
Exchangeable Na percentage	% of ECEC	< 15
Exchangeable Ca:Mg ratio	Ratio	3–9
Available phosphorus	mg/kg	
Acid soils method		50–150
Alkaline soils method		20–50
Available nitrogen (N as nitrate)	mg/kg	20–60

Part C. Example components for the soil supplier

The following table outlines suggested components that may likely meet the physical requirements of this specification. This is *not* part of the product specification. It is an example for the edification of the soil supplier of what might meet the product specification.

Example components (likely to meet the physical requirements of this specification):

Medium grade clean sand. Sandy loam soil or site soil	60–80% by volume	E.g. 7 parts washed
Composted soil conditioner conforming with	10–30% by volume	sand/2 part sandy
AS 4454	10% by volume	loam/1 part
		AS 4454 compost

Base level requirements for fertilisers (to be verified by laboratory testing and per agronomist's report):

Lime and/or dolomite	2 kg/m^3 at mixing
Balanced compound NPK turf starter fertiliser Minor	2.9 kg/100 m^2 after placement
and trace elements	300 g/m^3 at mixing

For the purposes of tendering, the contractor must allow for the inclusion of the above soil amendments, but the specific amendments required must be verified by laboratory testing and agronomist's recommendations.

See **Specifications G1, G2** and **G3** for validation, certification and test method specifications.

SPECIFICATION C3: SPORTS FIELD TURF

Part A. 'Fit-for-purpose' performance description

A sandy root-zone soil for construction of high-grade competitive sports fields. The physical specification allows for a minimum amount of silt and clay to provide cohesion and superior water-holding ability to the conventional USGA soil specification commonly used for tee and green construction.

Part B. Product specification (technical parameters)

Generally the soil must be free of 'unwanted material' and must meet all the require-ments of Tables 6.7 and 6.8. Where variations from these requirements occur refer to **Specifications G1** and **G2**.

> Certification and 'fit-for-purpose' statement to be submitted (with any soil amendments needed). Refer to **Specifications G1** and **G2** for the testing and certification requirements.

Table 6.7. Physical properties particle size analysis

Property	Units	Target range
2.0 mm (fine gravel)	% retained by mass	< 3
1.0 mm (very coarse sand)	% retained by mass	< 10
0.5 mm (coarse sand)	% retained by mass	10–30
0.25 mm (medium sand)	% retained by mass	30–50
0.1 mm (fine sand)	% retained by mass	20–30
0.05 (very fine sand)	% retained by mass	< 15 (max 20% combined vfs, Si + Cl)
0.002 mm (silt)	% retained by mass	< 6 (Si + clay combined 5–8%)
< 0.002 mm (clay)	% retained by mass	2–5
Large particles	% by mass	> 5 mm < 2%
Organic matter content	% w/w	2 to 5
Permeability*	mm/h	50–200
Water repellence	seconds	> 60

*ASTM F1815–1997

Table 6.8. Chemical properties

Property	Units	Target range
pH in water (1:5)	pH units	5.4–8.0
pH in CaCl2 (1:5)	pH units	5.2–7.5
Electrical conductivity (1:5)	dS/m	< 0.5
Exchangeable Na percentage	% of ECEC	< 15
Exchangeable Ca:Mg ratio	Ratio	3–9
Available phosphorus	mg/kg	
Acid soils method		50–150
Alkaline soils method		20–50
Available nitrogen (N as nitrate)	mg/kg	20–60

Part C. Example components for the soil supplier

The following table outlines suggested components that may likely meet the physical requirements of this specification. This is not part of the product specification. It is an example for the edification of the soil supplier of what might meet the product specification.

Example components (likely to meet the physical requirements of this specification):

Medium grade clean sand. Sandy loam soil or site soil	80–90% by volume	E.g. 8 parts washed
Composted soil conditioner conforming with	5–10% by volume	sand/2 parts sandy
AS 4454	5–10% by volume	loam/1 part
		AS 4454 compost

Base level requirements for fertilisers (to be verified by laboratory testing and per agronomist's report):

Lime and/or dolomite	2 kg/m^3 at mixing
Balanced compound NPK turf starter fertiliser Minor	2.9 kg/100 m^2 after placement
and trace elements	300 g/m^3 at mixing

For the purposes of tendering, the contractor must allow for the inclusion of the above soil amendments but the specific amendments required must be verified by laboratory testing and agronomist's recommendations.

See **Specifications G1**, **G2** and **G3** for validation, certification and test method specifications.

PART III D SOILS FOR GARDENS AND MASS PLANTING

How to use the Part III D specifications

There are four typical specification options for gardens and mass planting to choose from. These are:

- D1 Mass planting soil (for mainly low nutrient requirement plants).
- D2 Garden bed planting soil (standard garden soil).
- D3 High fertility display and vegetable production topsoil (premium garden soil).
- D4 Advanced tree and vault soils (for larger trees and trees in limited spaces).

If your design calls for garden beds on grade, select the typical Part III D specification as appropriate to your design intent and add all the parts of that specification (Parts A, B and C) to form part of your landscape specification document.

PDF templates of the soil specifications are available at <https://www.publish.csiro.au/book/8226>

Note: these specifications are typical only, not site specific, and may require specialist advice where site conditions or performance criteria is outside the parameters of these typical specifications.

These typical specifications have been provided for use by proficient and experienced landscape design professionals who are able to determine when these specifications are suitable. Where there is any doubt, consult with appropriate experienced professionals such as restoration ecologists, soil scientists or experienced landscape architects/consulting arborists.

SPECIFICATION D1: MASS PLANTING SOIL

Part A. 'Fit-for-purpose' performance description

A sandy loam to clay loam topsoil mix designed for mass planting of grasses, woody and herbaceous perennials that do not have very high nutrient requirements and is not subject to compaction by pedestrian or other traffic. The heavier textured soils in this specification may require the use of engineered solutions (drainage techniques) where excessive wetness is anticipated. Planting methods may vary and include direct seeding, tube and potted specimens up to 45 L pot size.

This planting specification can use site-won topsoil characterised according to **Specification A1**. Appropriate recycled soils may also be incorporated.

Choose from the following alternatives based upon the soil approach and design approach method:

Alkaline soils	Phosphorus-sensitive plants
☐ YES – Soils are alkaline or imported soils may be alkaline within the given alkaline range (refer to Table 6.10) ☐ NO – Soils are to be within the standard range of pH from Table 6.10	☐ YES – Phosphorus-sensitive plants are included in the design. Phosphorus level must be in the low P range from Table 6.10 ☐ NO – Phosphorus-tolerant plants have been chosen. Phosphorus levels must be within the standard range from Table 6.10

Note: if the above selections are not chosen, the landscape contractor/soil supplier must communicate with the landscape architect/specifier for determination.

Part B. Product specification (technical parameters)

Generally, the soil must be free of 'unwanted material' and must meet all the requirements of Tables 6.9 and 6.10. Where variations from these requirements occur refer to **Specifications G1** and **G2**.

> Certification and 'fit-for-purpose' statement to be submitted (with any soil amendments needed). Refer to **Specifications G1** and **G2** for the testing and certification requirements.

Table 6.9.   Physical properties

Property	Units	Target range
Texture, preferred range	n/a	Sandy loam to clay loam
Organic matter	% w/w	2–5
Permeability (Mc&J) or if texture is heavier than fine sandy clay loam, use 'estimated permeability'	mm/h	> 20. At 16 drop compaction level
Water repellence	seconds	> 60
Rock or foreign materials		
> 20 mm	% w/w	< 20

Table 6.10.   Chemical properties

Property	Units	Target range
pH in water (1:5) standard range	pH units	5.4–6.8
pH in CaCl$_2$ (1:5) standard range	pH units	5.2–6.5
pH in water (1:5) alkaline range	pH units	6.8–8.0
pH in CaCl$_2$ (1:5) alkaline range	pH units	6.5–7.5
Electrical conductivity (1:5)	dS/m	< 0.5
Phosphorus – P-tolerant/standard plants, acid soils method	mg/kg	30–100
Phosphorus – P-tolerant/standard plants, alkaline soils method	mg/kg	10–30
Phosphorus for P-sensitive plants, acid soils method	mg/kg	5–30
Phosphorus for P-sensitive plants, alkaline soils method	mg/kg	3–8
Exchangeable sodium (Na)	% of ECEC	< 15
Exchangeable potassium (K)	% of ECEC	3–10
Exchangeable calcium (Ca)	% of ECEC	Normal soil 60–80 Alkaline soil 70–90
Exchangeable magnesium (Mg)	% of CEC	15–25
Exchangeable aluminium (Al)	% of CEC	< 5
Exchangeable Ca:Mg ratio	Ratio	3–9
Available iron (Fe)	mg/kg	100–400
Available manganese (Mn)	mg/kg	25–100
Available zinc (Zn)	mg/kg	5–30
Available copper (Cu)	mg/kg	1–15
Available boron (B)	mg/kg	0.5–5
Available N (N as nitrate)	mg/kg	20–50

Part C. Example components for the soil supplier

The following table outlines suggested components that may likely meet the physical requirements of this specification. This is not part of the product specification. It is an example for
the edification of the soil supplier of what might meet the product specification.

Example components (likely to meet the physical requirements of this specification):

Sandy loam soil or site won topsoil	70–100% by volume	E.g. 8 parts washed sand/2 parts sandy loam/1 part AS 4454 compost
Composted soil conditioner conforming with AS 4454	0–30% by volume	

Base level requirements for fertilisers (to be verified by laboratory testing and per
agronomist's report):

Lime and/or dolomite	2 kg/m^3 at mixing
Balanced compound NPK turf starter fertiliser Minor and trace elements	0.5 kg/100 m^2 after placement 300 g/m^3 at mixing

For the purposes of tendering the contractor must allow for the inclusion of the above
soil amendments but the specific amendments required must be verified by laboratory
testing and agronomist's recommendations.

See **Specifications G1, G2** and **G3** for validation, certification and test method
specifications.

SPECIFICATION D2: GARDEN BED PLANTING SOIL

Part A. 'Fit-for-purpose' performance description

A sandy loam to clay loam topsoil mix designed for general purpose, on-grade landscape garden bed planting of grasses, woody and herbaceous annuals and perennials that have a high nutrient requirement for sustained optimum growth, and are not subject to compaction by pedestrian or other traffic.

The heavier textured soils in this specification may require the use of engineered solutions (drainage techniques) where excessive wetness is anticipated. Note that the organic soil variant should not be chosen for low P plantings and should not be used below 300 mm. Planting methods may vary and include direct seeding, tube and potted specimens up to 45 L pot size.

This planting specification may use site-won topsoil characterised according to **Specification A1**. Where site soils are *alkaline* (pH > 7.5 in water), the pH preferences of the planting list must be considered in the design process. Similarly, where salinity is identified as a potential issue, the salt tolerance of planting lists must be considered.

Choose from the following alternatives based on the soil approach and design approach method:

Alkaline soils	Phosphorus-sensitive plants
☐ YES – Soils are alkaline or imported soils may be alkaline within the given alkaline range (refer to Table 6.12)	☐ YES – Phosphorus-sensitive plants are included in the design. Phosphorus level must be in the low P range from Table 6.12
☐ NO – Soils are to be within the standard range of pH from Table 6.12	☐ NO – Phosphorus-tolerant plants have been chosen. Phosphorus levels must be within the standard range from Table 6.12

Note: if the above selections are not chosen, the landscape contractor/soil supplier must communicate with the landscape architect/specifier for determination.

Part B. Product specification (technical parameters)

Generally the soil must be free of 'unwanted material' and must meet all the requirements of Tables 6.11 and 6.12. Where variations from these requirements occur refer to **Specifications G1 and G2**.

> Certification and 'fit-for-purpose' statement to be submitted (with any soil amendments needed). Refer to **Specifications G1** and **G2** for the testing and certification requirements.

Part C. Example components for the soil supplier

The following table outlines suggested components that may likely meet the physical requirements of this specification. This is not part of the product specification. It is an example for the edification of the soil supplier of what might meet the product specification.

Example components (likely to meet the physical requirements of this specification):

Sandy loam soil or site won topsoil	70–90% by volume	E.g. 8 parts washed
Composted soil conditioner conforming with AS 4454	10–30% by volume	sand/2 parts sandy loam/1 part AS 4454
	30–60% by volume for organic soil variant	compost

Table 6.11. Physical properties

Property	Units	Target range
Texture, preferred range	n/a	Sandy loam to clay loam
Organic matter	% w/w	3–6
Organic matter (organic soil variant) dry weight basis (dwb)	% dwb	15–25
Permeability (Mc&J) or if texture is heavier than fine sandy clay loam, use 'estimated permeability'	mm/h	30–80. At 16 drop compaction level
Water repellence	seconds	> 60
Rock or foreign materials		
> 20 mm	% w/w	<20

Table 6.12. Chemical properties

Property	Units	Target range
pH in water (1:5) standard range	pH units	5.4–6.8
pH in CaCl$_2$ (1:5) standard range	pH units	5.2–6.5
pH in water (1:5) alkaline range	pH units	6.8–8.0
pH in CaCl$_2$ (1:5) alkaline range	pH units	6.5–7.5
Electrical conductivity (1:5)	dS/m	< 0.65
Phosphorus – P-tolerant/standard plants. Acid soils method	mg/kg	50–150
Phosphorus – P-tolerant/standard plants. Alkaline soils method	mg/kg	30–60
Phosphorus for P-sensitive plants, acid soils method	mg/kg	5–30
Phosphorus for P-sensitive plants, alkaline soils method	mg/kg	3–20
Exchangeable sodium (Na)	% of ECEC	< 15
Exchangeable potassium (K)	% of ECEC	5–10
Exchangeable calcium (Ca)	% of ECEC	Normal soil 60–80 Alkaline soil 70–90
Exchangeable magnesium (Mg)	% of CEC	15–25
Exchangeable aluminium (Al)	% of CEC	< 2
Exchangeable Ca:Mg ratio	Ratio	3–9
Available iron (Fe)	mg/kg	100–400
Available manganese (Mn)	mg/kg	25–100
Available zinc (Zn)	mg/kg	5–30
Available copper (Cu)	mg/kg	1–15
Available boron (B)	mg/kg	0.5–5
Available N (N as nitrate)	mg/kg	30–50

Base level requirements for fertilisers (to be verified by laboratory testing and per agronomist's report):

Lime and/or dolomite	2 kg/m^3 at mixing
Balanced compound NPK turf starter fertiliser Minor and trace elements	1.0 kg/100 m^2 after placement 300 g/m^3 at mixing

For the purposes of tendering, the contractor must allow for the inclusion of the above soil amendments, but the specific amendments required must be verified by laboratory testing and agronomist's recommendations.

See **Specifications G1**, **G2** and **G3** for validation, certification and test method specifications.

- -

SPECIFICATION D3: HIGH FERTILITY DISPLAY AND VEGETABLE PRODUCTION TOPSOIL

Part A. 'Fit-for-purpose' performance description

A high organic matter topsoil of an organic sandy loam to clay loam texture designed for on-grade, high-quality feature and display landscape bed planting of grasses, woody and herbaceous annuals and perennials that have a high nutrient requirement for sustained optimum growth, and is not subject to compaction by pedestrian or other traffic. Such installations typically have a *high turnover rate of 6 to 12 months*. The organic soil variant may be chosen for such installations. The organic soil variant, supported with a suitable fertiliser program, would be suited to fruit and vegetable production in urban farms.

The heavier textured soils in this specification may require the use of engineered solutions (drainage techniques) where excessive wetness is anticipated. Planting methods may vary and include direct seeding, tube and potted specimens up to 45 L.

This planting specification may use site-won topsoil characterised according to **Specification A1**. Where site soils are *alkaline* (pH > 7.5 in water), the pH preferences of the planting list must be considered in the design process. Similarly, where *salinity* is identified as a potential issue, the salt tolerance of planting lists must be considered.

Choose from the following alternatives based on the soil approach and design approach method:

Alkaline soils	Phosphorus-sensitive plants
☐ YES – Soils are alkaline or imported soils may be alkaline within the given alkaline range (refer to Table 6.14) ☐ NO – Soils are to be within the standard range of pH from Table 6.14	☐ YES – Phosphorus-sensitive plants are included in the design. Phosphorus level must be in the low P range from Table 6.14 ☐ NO – Phosphorus-tolerant plants have been chosen. Phosphorus levels must be within the standard range from Table 6.14

Note: if the above selections are not chosen, the landscape contractor/soil supplier must communicate with the landscape architect/specifier for determination.

Part B. Product specification (technical parameters)

Generally the soil must be free of 'unwanted material' and must meet all the requirements of Tables 6.13 and 6.14. Where variations from these requirements occur, refer to **Specifications G1** and **G2**.

Certification and 'fit-for-purpose' statement to be submitted (with any soil amendments needed). Refer to **Specifications G1** and **G2** for the testing and certification requirements.

Part C. Example components for the soil supplier

The following table outlines suggested components that may likely meet the physical requirements of this specification. This is *not* part of the product specification. It is an example for the edification of the soil supplier of what might meet the product specification.

Example components (likely to meet the physical requirements of this specification):

Sandy loam soil or site won topsoil	40–60% by volume
Composted soil conditioner conforming with AS 4454	20–30% by volume 30–50% organic soil variant
Composted manure	10–30% by volume

Table 6.13.   Physical properties

Property	Units	Target range
Texture, preferred range	n/a	Sandy loam to clay loam
Organic matter	% w/w	3–6
Organic matter (organic soil variant) dry weight basis (dwb)	% dwb	15–25
Permeability (Mc&J) or if texture is heavier than fine sandy clay loam, use 'estimated permeability'	mm/h	> 50. At 16 drop compaction level
Water repellence	seconds	> 60
Rock or foreign materials		
> 20 mm	% w/w	< 20

Table 6.14.   Chemical properties

Property	Units	Target range
pH in water (1:5) standard range	pH units	5.4–6.8
pH in CaCl2 (1:5) standard range	pH units	5.2–6.5
pH in water (1:5) alkaline range	pH units	6.8–8.0
pH in CaCl2 (1:5) alkaline range	pH units	6.5–7.5
Electrical conductivity (1:5)	dS/m	< 0.75
Phosphorus – P-tolerant/standard plants. Acid soils method	mg/kg	80–150
Phosphorus – P-tolerant/standard plants. Alkaline method	mg/kg	40–80
Phosphorus for P-sensitive plants, acid soils method	mg/kg	< 30
Phosphorus for P-sensitive plants, alkaline soils method	mg/kg	< 20
Exchangeable sodium (Na)	% of ECEC	< 15
Exchangeable potassium (K)	% of ECEC	5–15
Exchangeable calcium (Ca) method 18F1 or 15A1 in alkaline soils	% of ECEC	Normal soil 60–80 Alkaline soil 70–90
Exchangeable magnesium (Mg)	% of CEC	15–25
Exchangeable aluminium (Al)	% of CEC	< 2
Exchangeable Ca:Mg ratio	Ratio	3–9
Available iron (Fe)	mg/kg	100–400
Available manganese (Mn)	mg/kg	25–100
Available zinc (Zn)	mg/kg	5–30
Available copper (Cu)	mg/kg	1–15
Available boron (B)	mg/kg	0.5–5
Available N (N as nitrate)	mg/kg	> 50

Base level requirements for fertilisers (to be verified by laboratory testing and per agronomist's report):

Lime and/or dolomite	2 kg/m^3 at mixing
Balanced compound NPK turf starter fertiliser Minor and trace elements	1.0 kg/100 m^2 after placement 300 g/m^3 at mixing

For the purposes of tendering, the contractor must allow for the inclusion of the above soil amendments, but the specific amendments required must be verified by laboratory testing and agronomist recommendations.

See **Specifications G1, G2** and **G3** for validation, certification and test method specifications.

High fertility display and vegetable production soil fertiliser applications

The suggested fertilisers listed in the example are expected to last 3–6 months of sustained growth and performance. A suitable fertiliser and organic matter maintenance program may be required after this period depending on the nature of the garden type.

Note: generally when planting advanced trees (with root balls deeper than 400 mm or pots larger than 45 L), it is advisable to use a free-draining sandy backfill medium that is low in organic matter below 300 mm to reduce the risk of waterlogging and oxygen depletion. Structural cells may increase the oxygen levels at depth. The topsoil (A Horizon to 300–400 mm depth) and other areas without advanced tree plantings may comprise **Specification D1**, **D2** or **D3** as appropriate.

SPECIFICATION D4: ADVANCED TREE AND VAULT SUBSOILS

Part A. 'Fit-for-purpose' performance description

A sandy, well-drained medium with low organic matter for backfilling below 300 mm from the surface in larger potted specimens over 45 L or 400 mm depth of root ball, semi-advanced, advanced and super-advanced tree planting. The specification may use a small proportion of site won topsoil or subsoil, provided the organic matter upper limit is not exceeded. Above 300 mm preferably use **Specification D2**.

Where site soils or imported soils are *alkaline* (pH > 7.5 in water), the pH preferences of the planting list must be considered in the design process.

Choose from the following alternatives based upon the soil approach and design approach method:

Alkaline soils	Phosphorus-sensitive plants
☐ YES – Soils are alkaline or imported soils may be alkaline within the given alkaline range (refer to Table 6.16) ☐ NO – Soils are to be within the standard range of pH from Table 6.16	☐ YES – Phosphorus-sensitive plants are included in the design. Phosphorus level must be in the low P range from Table 6.16 ☐ NO – Phosphorus-tolerant plants have been chosen. Phosphorus levels must be within the standard range from Table 6.16

Note: if the above selections are not chosen, the landscape contractor/soil supplier must communicate with the landscape architect/specifier for determination.

Part B. Product specification (technical parameters)

Generally, the soil must be free of 'unwanted material' and must meet all the requirements of Tables 6.15 and 6.16. Where variations from these requirements occur refer to **Specifications G1** and **G2**.

> Certification and 'fit-for-purpose' statement to be submitted (with any soil amendments needed). Refer to **Specifications G1** and **G2** for the testing and certification requirements.

Table 6.15.   Physical properties

Property	Units	Target range
Texture, preferred range	n/a	Loamy sand to Sandy loam
Organic matter	% w/w	< 5
Permeability (Mc&J) or if texture is heavier than fine sandy clay loam, use 'estimated permeability'	mm/h	> 50. At 16 drop compaction level
Water repellence	seconds	> 60
Rock or foreign materials		
> 20 mm	% w/w	< 20

Table 6.16.   Chemical properties

Property	Units	Target range
pH in water (1:5) standard range	pH units	5.4–6.8
pH in CaCl$_2$ (1:5) standard range	pH units	5.2–6.5
pH in water (1:5) alkaline range	pH units	6.8–8.0
pH in CaCl$_2$ (1:5) alkaline range	pH units	6.5–7.5
Electrical conductivity (1:5)	dS/m	< 0.5
Phosphorus – P-tolerant/standard plants. Acid soils method	mg/kg	30–80

Property	Units	Target range
Phosphorus – P-tolerant/standard plants. Alkaline method 9B1 or 9C1	mg/kg	10–30
Phosphorus for P-sensitive plants, acid soils method	mg/kg	5–30
Phosphorus for P-sensitive plants, alkaline soils method	mg/kg	2–20
Exchangeable sodium (Na)	% of ECEC	< 5
Exchangeable potassium (K)	% of ECEC	3–10
Exchangeable calcium (Ca)	% of ECEC	Normal soil 60–80 Alkaline soil 70–90
Exchangeable magnesium (Mg)	% of CEC	15–25
Exchangeable aluminium (Al)	% of CEC	< 5
Exchangeable Ca:Mg ratio	Ratio	3–9
Available iron (Fe)	mg/kg	100–400
Available manganese (Mn)	mg/kg	25–100
Available zinc (Zn)	mg/kg	5–30
Available copper (Cu)	mg/kg	1–15
Available boron (B)	mg/kg	0.5–5
Available N (N as nitrate)	mg/kg	20–40

Part C. Example components for the soil supplier

The following table outlines suggested components that may likely meet the physical requirements of this specification. This is *not* part of the product specification. It is an example for the edification of the soil supplier of what might meet the product specification.

Example components (likely to meet the physical requirements of this specification):

Sandy loam soil	60–80% by volume
On-site clay loam or clay topsoil or subsoil	20–30% by volume
Composted soil conditioner conforming with AS 4454	< 10% by volume

Base level requirements for fertilisers (to be verified by laboratory testing and per agronomist's report):

Lime and/or dolomite	2 kg/m^3 at mixing
or	
Gypsum	2 kg/m^3 at mixing

For the purposes of tendering, the contractor must allow for the inclusion of the above soil amendments, but the specific amendments required must be verified by laboratory testing and agronomist's recommendations.

See **Specifications G1**, **G2** and **G3** for validation, certification and test method specifications.

PART III E ON SLAB MEDIA

How to use the Part III E specifications

There are four typical specification options for on-slab media to choose from. These are:

- E1 On slab growing media 'A' horizon (for low grasses and groundcovers).
- E2 On slab growing media 'B' horizon (for low shrubs and plants).
- E3 Low-density container and green roof (for lightweight requirements).
- E4 Ultra lightweight growing media 'A' only horizon (for extensive rooftops with shallow growing media profiles).

If your design calls for landscapes that are 'on-slab', 'above ground' or 'on podium', select the typical Part III E Specification as appropriate to your design intent and add all the parts of that specification (Parts A, B and C) to form part of your landscape specification document.

Note: E1 and E4 are considered an A horizon growing media and E2 is a subsoil. It is likely that both E1 and E2 will both need to be used where profile horizon structure is an A/B Profile (i.e. where growing media is greater than 300 mm in depth). For more information on profile horizon structures, go to **Specification B4**.

Specification E4 is for an extensive rooftop application, typically a shallow and broad 'extensive' area to support groundcovers, generally seldom accessed, except for very minimal maintenance and typically 150–300 mm in depth with only one profile horizon.

PDF templates of the soil specifications are available at <https://www.publish.csiro.au/book/8226>.

Note: these specifications are typical only, not site specific, and may require specialist advice where site conditions or performance criteria is outside the parameters of these typical specifications.

These typical specifications have been provided for use by proficient and experienced landscape design professionals who are able to determine when these specifications are suitable. Where there is any doubt, consult with appropriate experienced professionals such as restoration ecologists, soil scientists or experienced landscape architects/consulting arborists.

- -

SPECIFICATION E1: ON SLAB SOIL MEDIA 'A' HORIZON

Part A. 'Fit-for-purpose' performance description

The specification describes the formulation of an open, granular, well-drained growing media with a saturated density of between 1800 kg/m m^3 (1.8 kg/L) and 2400 kg/m^3 (2.4 kg/L) for use in on-slab applications, including green roofs with an expectation of longevity. It is a topsoil formulation to be used in the surface 300 mm of all on-slab installations including planter boxes, containers and garden beds.

In order to maintain structure and porosity over extended periods, and to avoid slumping and volume loss over time, the formulation must employ low-density mineral components such as ash, perlite, scoria, pumice and diatomaceous earth, or artificial components such as urea formaldehyde and Styrofoam^{TM}. Physically, the media has the properties of a potting media and is assessed using the methodology of AS 3743. Where a subsoil is also specified, use the 'B' horizon profile **Specification E2** as well as **Specification E1**.

> Certification and 'fit-for-purpose' statement to be submitted (with any soil amendments needed). Refer to **Specifications G1** and **G2** for the testing and certification requirements.

Choose from the following alternatives based upon the soil approach and design approach method:

Phosphorus-sensitive plants
☐ YES – Phosphorus-sensitive plants are included in the design. Phosphorus level must be in the low P range (refer to Table 6.18)
☐ NO – Phosphorus-tolerant plants have been chosen. Phosphorus levels must be within the standard range from Table 6.18

Note: if the above selections are not chosen, the landscape contractor/soil supplier must communicate with the landscape architect/specifier for determination.

Part B. Product specification (technical parameters)

Generally, the soil must be free of 'unwanted material' and must meet all the requirements in Tables 6.17 and 6.18, *AS 3743 Potting Mixes* and the specified requirements of AS 4419. However, compliance with AS 3743 does not demonstrate compliance with this specification. Where the requirements of this specification and AS 3743 conflict, properties specified here must take precedence.

Table 6.17.   Physical properties

Property	Units	Optimum target range
Texture, preferred range	n/a	Gravelly loamy sand to organic sandy loam
Air-filled porosity (AFP)	%	10–20 A Horizon (300 mm maximum E1 depth).
Water-holding capacity (WHC)	%	40–50
Permeability (Mc&J) or if texture is heavier than fine sandy clay loam, use 'estimated permeability'	mm/h	> 100. At 16 drop compaction level
Organic matter (OM) or Total Organic Content (TOC)	% w/w	< 15 (especially where P sensitive) Or 15–25 where high maintenance input or short lifespan and where P tolerant species.
Water repellence	seconds	> 60
Saturated repacked density	kg/m^3	1800–2400

Table 6.18.   Chemical properties

C2. Chemical properties	Units	Target range
pH in water (1:1.5) standard range	pH units	5.4–6.8
Electrical conductivity (1:1.5)	dS/m	< 2.2
Chloride (1:1.5)	mg/L	≤ 200
Ammonium-N (NH$_4$ 1:1.5)	mg/L	≤ 100
Ammonium-N + nitrate-N (NH$_4$ + NO$_3$)	mg/L	50–100
Nitrogen draw-down index	–	≥ 0.7
Toxicity index	mm	≥ 70
Phosphorus – P standard range	mg/L	8–40
Low phosphorus – P (P-sensitive plants)	mg/L	< 3
Potassium (K)	mg/L	50–250
Sulphate (SO$_4$)	mg/L	≥ 40
Calcium (Ca)	mg/L	≥ 80
Magnesium (Mg)	mg/L	≥ 15
Ca:Mg ratio	Ratio	1.5–10
K:Mg ratio	Ratio	1–7
Sodium (Na)	mg/L	< 150
Iron (Fe)	mg/L	≥ 35
Copper (Cu)	mg/L	0.4–15
Zinc (Zn)	mg/L	0.3–10
Manganese (Mn)	mg/L	1–15
Boron (B)	mg/L	0.02–0.65

Use AS 3743 unless otherwise stated. See **Specification G3** for other test methods.

Part C. Example components for the soil supplier

The following table outlines suggested components that may likely meet the physical requirements of this specification. This is not part of the product specification. It is an example for the edification of the soil supplier of what might meet the product specification.

Example components (likely to meet the physical requirements of this specification):

Sandy loam soil or site won topsoil	20–40% by volume
Horticultural ash, perlite, or similar lightweight low-density mineral matter or mixtures of these	30–60% by volume
	20–30% by volume
Composted soil conditioner conforming with AS 4454	

Note: when ash is used an acceptable exceedance of the organic matter target may occur due to the presence of inert carbon.

Base level requirements for fertilisers (to be verified by laboratory testing and per agronomist's report):

Lime and/or dolomite	2 kg/m^3 at mixing
Balanced compound NPK turf starter fertiliser	3.0 kg/100 m^2 after placement
Minor and trace elements	300 g/m^3 at mixing

For the purposes of tendering, the contractor must allow for the inclusion of the above soil amendments, but the specific amendments required must be verified by laboratory testing and agronomist's recommendations.

The *suggested fertilisers are expected to last 3–6 months* of sustained growth. A suitable fertiliser (e.g. controlled slow release) and organic matter maintenance program may be required after this period, depending on the design intent.

See **Specifications G1, G2** and **G3** for validation, certification and test method specifications.

SPECIFICATION E2: ON SLAB SOIL MEDIA 'B' HORIZON

Part A. 'Fit-for-purpose' performance description

The specification describes the formulation of an open, granular well-drained growing media with a saturated density of between 1800 kg/m m^3 (1.8 kg/L) and 2400 kg/m^3 (2.4 kg/L) for use in on-slab applications with an expectation of longevity to be used as a subsoil below 300 mm of all on-slab installations, including planter boxes, containers and garden beds. For the upper/above 300 mm 'A' horizon use **Specification E1**.

In order to maintain structure and porosity over extended periods, and to avoid slumping and volume loss over time, the formulation must employ low-density mineral components such as ash, perlite, scoria, pumice and diatomaceous earth, or artificial components such as urea formaldehyde and Styrofoam^{TM}.

Physically the media has the properties of a potting media and is assessed using the methodology of AS 3743.

> Certification and 'fit-for-purpose' statement to be submitted (with any soil amendments needed). Refer to Specifications **G1**, **G2** and **G3** for the testing and certification requirements.

Choose from the following alternatives based upon the soil approach and design approach method:

Phosphorus-sensitive plants
☐ YES – Phosphorus-sensitive plants are included in the design. Phosphorus level must be in the low P range from Table 6.20
☐ NO – Phosphorus-tolerant plants have been chosen. Phosphorus levels must be within the standard range from Table 6.20

Note: if the above selections are not chosen, the landscape contractor/soil supplier must communicate with the landscape architect/specifier for determination.

Part B. Product specification (technical parameters)

Generally, the soil must be free of 'unwanted material' and must meet all the requirements of AS 3743 and the specified requirements of AS 4419. However, compliance with AS 3743 does not demonstrate compliance with this specification. Where the requirements of this specification and AS 3743 conflict, properties specified in Table 6.19 must take precedence.

Table 6.19.   Physical properties

Property	Units	Optimum target range
Texture, preferred range	n/a	Gravelly loamy sand to organic sandy loam
Air-filled porosity (AFP)	%	≥ 10 in top 300 mm (A Horizon) 5–10 if in B Horizon up 800 mm total soil depth, or 5–8 in B Horizon between 800–1200 mm total soil depth
Water-holding capacity (WHC)	%	40–50
Permeability (Mc&J) or if texture is heavier than fine sandy clay loam, use 'estimated permeability'	mm/h	100–200. At 16 drop compaction level

Property	Units	Optimum target range
Organic matter (OM) % water weight	% w/w	5–20
Water repellence	seconds	> 60
Saturated repacked density	kg/m³	1800–2400

Table 6.20.   Chemical properties

Property	Units	Target range
pH in water (1:1.5) standard range	pH units	5.4–6.8
Electrical conductivity (1:1.5)	dS/m	< 2.2
Chloride (1:1.5)	mg/L	≤ 200
Ammonium-N (NH$_4$ 1:1.5)	mg/L	≤ 100
Ammonium-N + nitrate-N (NH$_4$ + NO$_3$ 1:1.5)	mg/L	50–130
Nitrogen draw-down index	–	≥ 0.7
Toxicity index	mm	≥ 70
Phosphorus – P	mg/L	8–40
Low phosphorus – P-sensitive plants	mg/L	< 3
Potassium (K)	mg/L	50–250
Sulphate (SO$_4$)	mg/L	≥ 40
Calcium (Ca)	mg/L	≥ 80
Magnesium (Mg)	mg/L	≥ 15
Ca:Mg ratio	Ratio	1.5–10
K:Mg ratio	Ratio	1–7
Sodium (Na)	mg/L	< 150
Iron (Fe)	mg/L	≥ 35
Copper (Cu)	mg/L	0.4–15
Zinc (Zn)	mg/L	0.3–10
Manganese (Mn)	mg/L	1–15
Boron (B)	mg/L	0.02–0.65

Use AS 3743 unless otherwise stated. See **Specification G3** for other test methods.

Part C. Example components for the soil supplier

The following table outlines suggested components that may likely meet the physical requirements of this specification. This is not part of the product specification. It is an example for the edification of the soil supplier of what might meet the product specification.

Example components (likely to meet the physical requirements of this specification):

Sandy loam soil or site won topsoil	10–30% by volume
Horticultural ash, perlite, or similar lightweight low-density mineral matter or mixtures of these	30–50% by volume
Composted 10 mm pine bark	20–40% by volume
Composted soil conditioner conforming with AS 4454	< 20% by volume

Note: when ash is used an acceptable exceedance of the organic matter target may occur due to the presence of inert carbon.

Base level requirements for fertilisers (to be verified by laboratory testing and per agronomist's report):

Lime and/or dolomite	2 kg/m^3 at mixing
Balanced compound NPK turf starter fertiliser Minor and trace elements	3.0 kg/100 m^2 after placement
	300 g/m^3 at mixing

For the purposes of tendering, the contractor must allow for the inclusion of the above soil amendments, but the specific amendments required must be verified by laboratory testing and agronomist's recommendations.

The *suggested fertilisers are expected to last 3–6 months* of sustained growth. A suitable fertiliser (e.g. controlled slow release) and organic matter maintenance program may be required after this period, depending on the design intent.

See **Specifications G1**, **G2** and **G3** for validation, certification and test method specifications.

SPECIFICATION E3: LOW-DENSITY CONTAINER AND GREEN ROOF

Part A. 'Fit-for-purpose' performance description

This specification describes the formulation of lightweight growing media with a saturated density between 1100 kg/ m^3 (1.1 kg/L) and 1800 kg/m^3 (1.8 kg/L), for use in plant containers, planter boxes and on-slab applications including green roofs with an expectation of longevity. Due to the high porosity of the mix, it does not require the use of a low organic matter on slab subsoil medium below 300 mm and may be used in deep containers.

In order to maintain structure and porosity over extended periods, and to avoid slumping and volume loss over time, the formulation must employ low-density mineral components such as ash, perlite, scoria, pumice and diatomaceous earth, or artificial components such as urea formaldehyde and Styrofoam^{TM}.

Physically the media has the properties of a potting media and is assessed using the methodology of AS 3743.

> Certification and 'fit-for-purpose' statement to be submitted (with any soil amendments needed). Refer to **Specifications G1** and **G2** for the testing and certification requirements.

Choose from the following alternatives based upon the soil approach and design approach method:

Phosphorus-sensitive plants
☐ YES – Phosphorus-sensitive plants are included in the design. Phosphorus level must be in the low P range (refer to Table 6.22)
☐ NO – Phosphorus-tolerant plants have been chosen. Phosphorus levels must be within the standard range from Table 6.22

Note: if the above selections are not chosen, the landscape contractor/soil supplier must communicate with the landscape architect/specifier for determination.

Part B. Product specification (technical parameters)

Generally, the soil must be free of 'unwanted material' and must meet all the requirements in Tables 6.21 and 6.22, AS 3743 and the specified requirements of *AS 4419 Landscape Soils*. However, compliance with AS 3743 does not demonstrate compliance with this specification. Where the requirements of this specification and AS 3743 conflict, properties specified here must take precedence.

Use AS 3743 unless otherwise stated. See **Specification G1** for other test methods.

Table 6.21.   Physical properties

Property	Units	Target range
Texture, preferred range	n/a	Gravelly loamy sand to organic sandy loam
Air-filled porosity (AFP)	%	10–15
Water-holding capacity	%	45–55
Permeability (Mc&J) or if texture is heavier than fine sandy clay loam, use 'estimated permeability'	mm/h	100–200. At 16 drop compaction level
Organic matter	% w/w	< 25
Water repellence	seconds	> 60
Saturated repacked density	kg/m^3	1300–1800

Table 6.22.   Chemical properties

Property	Units	Target range
pH in water (1:1.5) standard range	pH units	5.4–6.8
Electrical conductivity (1:1.5)	dS/m	< 2.2
Chloride (1:1.5)	mg/L	≤ 200
Ammonium-N (NH$_4$ 1:1.5)	mg/L	≤ 100
Ammonium-N + nitrate-N (NH$_4$ + NO$_3$ 1:1.5)	mg/L	50–100
Nitrogen draw-down index	–	≥ 0.7
Toxicity index	mm	≥ 70
Phosphorus (P) standard range	mg/L	8–40
Low phosphorus P-sensitive plants	mg/L	< 3
Potassium (K)	mg/L	50–250
Sulphate (SO$_4$)	mg/L	≥ 40
Calcium (Ca)	mg/L	≥ 80
Magnesium (Mg)	mg/L	≥ 15
Ca:Mg ratio	Ratio	1.5–10
K:Mg ratio	Ratio	1–7
Sodium (Na)	mg/L	< 150
Iron (Fe)	mg/L	≥ 35
Copper (Cu)	mg/L	0.4–15
Zinc (Zn)	mg/L	0.3–10
Manganese (Mn)	mg/L	1–15
Boron (B)	mg/L	0.02–0.65

Part C. Example components for the soil supplier

The following table outlines suggested components that may likely meet the physical requirements of this specification. This is not part of the product specification. It is an example for the edification of the soil supplier of what might meet the product specification.

Example components likely to meet the physical requirements of this specification:

Sandy or sandy loam soil	< 20% by volume
Horticultural ash, perlite, or similar lightweight low-density mineral matter or mixtures of these	30–60% by volume
Composted 10 mm pine bark	20–30% by volume
Composted soil conditioner conforming with AS 4454	10–30% by volume

Note: when ash is used an acceptable exceedance of the organic matter target may occur due to the presence of inert carbon.

Base level requirements for fertilisers (to be verified by laboratory testing and per agronomist's report):

Lime and/or dolomite	2 kg/m^3 at mixing
Balanced compound NPK turf starter fertiliser Minor and trace elements	3.0 kg/100 m^2 after placement 300 g/m^3 at mixing

For the purposes of tendering, the contractor must allow for the inclusion of the above soil amendments, but the specific amendments required must be verified by laboratory testing and agronomist's recommendations.

The *suggested fertilisers are expected to last 3–6 months* of sustained growth. A suitable fertiliser (e.g. controlled slow release) and organic matter maintenance program may be required after this period, depending on the design intent.

See **Specifications G1, G2** and **G3** for validation, certification and test method specifications.

SPECIFICATION E4: ULTRA LIGHTWEIGHT GROWING MEDIA 'A' ONLY HORIZON (FOR EXTENSIVE ROOFTOPS WITH SHALLOW GROWING MEDIA PROFILES)

Part A. 'Fit-for-purpose' performance description

This specification is for extensive rooftop application, typically a shallow and broad 'extensive' area to support groundcovers, generally seldom accessed, except for very minimal maintenance and typically 150–300 mm in depth with only one profile horizon. This shallow media is not suited for woody shrubs or larger shrubs or trees.

The lightweight saturated density is between 1100 kg/m^3 (1.1 kg/L) and 1300 kg/m^3 (1.3 kg/L). Generally no more than 20–30% v/v of organic components should be used to avoid shrinkage. Note: To design super-lightweight mixes with a saturated density less than 1100 kg/m^3 requires the use of synthetic artificial materials such as polystyrene and urea formaldehyde foams.

Physically the media has the properties of a potting media and is assessed using the methodology of AS 3743.

Choose from the following alternatives based upon the soil approach and design approach method:

Phosphorus-sensitive plants
☐ YES – Phosphorus-sensitive plants are included in the design. Phosphorus level must be in the low P range (refer to Table 6.24)
☐ NO – Phosphorus-tolerant plants have been chosen. Phosphorus levels must be within the standard range from Table 6.24

Note: if the above selections are not chosen, the landscape contractor/soil supplier must communicate with the landscape architect/specifier for determination.

Part B. Product specification (technical parameters)

Generally, the soil must be free of 'unwanted material' and must meet all the requirements in Tables 6.23 and 6.24, AS 3743 and the specified requirements of *AS 4419 Landscape Soils*. However, compliance with AS 3743 does not demonstrate compliance with this specification. Where the requirements of this specification and AS 3743 conflict, properties specified here must take precedence. Use AS 3743 unless otherwise stated.

> Certification and 'fit-for-purpose' statement to be submitted (with any soil amendments needed). Refer to **Specifications G1** and **G2** for the testing and certification requirements.

Table 6.23.   Physical properties

Property	Units	Target/range
Texture, preferred range	n/a	Gravelly sandy clay loam
Air-filled porosity	%	13–30
Water-holding capacity	%	30–40
Permeability (Mc&J) or if texture is heavier than fine sandy clay loam, use 'estimated permeability'	mm/h	20–80. At 16 drop compaction level
Organic matter	% w/w	< 25
Water repellence	seconds	> 60
Saturated repacked density	kg/m^3	1100–1300

Table 6.24. Chemical properties

Property	Units	Target range
pH in water (1:1.5) standard range	pH units	5.4–6.8
Electrical conductivity (1:1.5)	dS/m	< 2.2
Chloride (1:1.5)	mg/L	≤ 200
Ammonium-N (NH_4 1:1.5)	mg/L	≤ 100
Ammonium-N + nitrate-N (NH_4 + NO_3 1:1.5)	mg/L	50–80
Nitrogen draw-down index	–	≥ 0.7
Toxicity index	mm	≥ 70
Phosphorus (P) standard range	mg/L	8–40
Low phosphorus P-sensitive plants	mg/L	< 3
Potassium (K)	mg/L	50–250
Sulphate (SO_4)	mg/L	≥ 40
Calcium (Ca)	mg/L	≥ 80
Magnesium (Mg)	mg/L	≥ 15
Ca:Mg ratio	Ratio	1.5–10
K:Mg ratio	Ratio	1–7
Sodium (Na)	mg/L	< 150
Iron (Fe)	mg/L	≥ 35
Copper (Cu)	mg/L	0.4–15
Zinc (Zn)	mg/L	0.3–10
Manganese (Mn)	mg/L	1–15
Boron (B)	mg/L	0.02–0.65

Part C. Example components for the soil supplier

The following table outlines suggested components that may likely meet the physical requirements of this specification. This is not part of the product specification. It is an example for the edification of the soil supplier of what might meet the product specification.

Example components (likely to meet the physical requirements of this specification):

Coarse/super coarse perlite (2–8 mm)	20–40% v/v
Super fine perlite (500 micron – 2 mm)	10–20% v/v
Horticultural ashes – 20 mm	40–60% v/v
5 mm composted pine bark fines or coconut coir	10–20% v/v
10 mm minus (screened) low phosphorus compost (compliant to AS 4454)	10–20% v/v

Note: when ash is used an acceptable exceedance of the organic matter target may occur due to the presence of inert carbon.

Base level requirements for fertilisers (to be verified by laboratory testing and per agronomist's report)

Lime and/or dolomite	2 kg/m^3 at mixing
Balanced compound NPK controlled release fertiliser	300 g/100 m^2 after placement or
Minor and trace elements	4 kg/m^3 during mixing
	300 g/m^3 at mixing

For the purposes of tendering, the contractor must allow for the inclusion of the above soil amendments, but the specific amendments required must be verified by laboratory testing and agronomist's recommendations.

The *suggested fertilisers are expected to last 3–12 months* of sustained growth depending on the control release fertiliser (CRF) chosen. A suitable fertiliser (e.g. controlled slow release) and organic matter maintenance program may be required after this period, depending on the design intent.

See **Specifications G1, G2** and **G3** for validation, certification and test method specifications.

PART III F SPECIALIST SOILS

How to use the Part III F specifications

There are three typical specification options for specialist soils to choose from. These are:

- F1 Structural support soils (SSS) (in vaults/where surface loads are required).
- F2 Raingardens, biofiltration, and stormwater filtration soils.
- F3 Wetland soils.

If your design calls for landscapes that are applicable to the listed specification applications, *select* the typical Part III F specification as appropriate to your design intent and add *all* the parts of that specification (Parts A, B and C) to form part of your landscape specification document.

Note: make clear the areas in the landscape that require the particular specialist soils.

PDF templates of the soil specifications are available at: <https://www.publish.csiro.au/book/8226>.

Note: these specifications are typical only, not site specific, and may require specialist advice where site conditions or performance criteria is outside the parameters of these typical specifications.

These typical specifications have been provided for use by proficient and experienced landscape design professionals who are able to determine when these specifications are suitable. Where there is any doubt, consult with appropriate experienced professionals such as restoration ecologists, soil scientists or experienced landscape architects/consulting arborists.

✂ -

SPECIFICATION F1: STRUCTURAL SUPPORT SOIL

Part A. 'Fit-for-purpose' performance description

The specification describes the formulation of a structural support soil (SSS) for tree planting in urbanised environments. SSS is designed to form a basement for engineered structures such as roads, pavements and kerbing, while also providing rooting volume for tree roots. Due to the high void space, it will permit root growth through the medium and also help distribute root pressures over a wider section of pavement, reducing or delaying pavement heaving by roots. The size of the aggregates or stone fraction determines how large the roots can grow before heaving occurs.

SSS is a two-part system comprised of a stone lattice for strength and structural support (load bearing) and filler soil to service the horticultural needs. The stone lattice provides structural stability through stone-to-stone contact, while also providing interconnected voids for root penetration, air and water movement. The system is engineered to maintain a high degree of porosity after installation and compaction. The intention is to 'suspend' the horticultural soil component of the blend between stones, which come together during compaction, producing a load-bearing, compacted stone lattice with uncompacted soil in the voids.

The ratio of filler soil to aggregate is the major consideration for achieving the engineering and horticultural objective. Thus the 'aggregate', 'filler soil' and blending ratio of the two need to be specified and carefully validated. Generally, it will be an amount of filler soil equal to half the void space of the compacted aggregate. Assuming the aggregate has a void space of 40%, it will be 10 parts aggregate by volume to 2 parts filler soil.

> *Important note:* the total volume of SSS is determined by the volume of aggregate as adding filler soil does not increase the overall volume. Tree soil volume estimations must factor this into calculations for recommending soil volume.

Transport and placement of SSS

SSS must be a uniformly blended mixture of aggregate and filler soil and is prone to segregation during handling at the source and during transport. Particular care must be taken to ensure that all structural soil is thoroughly homogenised before placement and compaction. To assist this, and to prevent segregation, ensure that the mixture remains moist at all times during mixing, transport, storage and placement.

Part B. Product specification (technical parameters): filler soil

The criteria provided in Tables 6.25 and 6.26 must be applied to the filler soil component of the SSS blend. In addition to the performance specification listed below, the filler soil component must be a clay loam or similar texture and be of uniform composition without admixture of subsoil and must be free of 'unwanted material'. It must be free of stone and gravel greater than 8 mm and be free from toxic substances harmful to plant growth.

> Certification and 'fit-for-purpose' statement to be submitted (with any soil amendments needed). Refer to **Specifications G1** and **G2** for the testing and certification requirements.

Tables 6.25 to 6.27 provide acceptable chemical and physical properties for the rock or aggregate components of the SSS blend.

Table 6.25.   Physical properties

Property	Units	Target range
Texture, preferred range	n/a	Loam to clay loam
Organic matter	% w/w	3–8
Water repellence	seconds	> 60
Gravel > 4 mm	% w/w	< 2

The following assessment criteria apply to the nominal 63 mm aggregate to be used in the SSS blend. The aggregate must be a free-draining granular material capable of sustaining the anticipated load bearing requirements of the pavement and must be free of 'unwanted material'. As a guide, an aggregate that conforms to the requirements of AS 2758.7 (1996) for Class L 60 mm railway ballast is likely to possess the desired properties.

Table 6.26.   Chemical properties

Property	Units	Target range
pH in water (1:5) standard range	pH units	5.4–6.8
pH in $CaCl_2$ (1:5) standard range	pH units	5.2–6.5
Electrical conductivity (1:5)	dS/m	< 0.5
Phosphorus (P)	mg/kg	30–100
Exchangeable sodium (Na)	% of ECEC	< 5
Exchangeable potassium (K)	% of ECEC	3–10
Exchangeable calcium (Ca)	% of ECEC	60–80
Exchangeable magnesium (Mg)	% of CEC	15–25
Exchangeable aluminium (Al)	% of CEC	< 5
Exchangeable Ca:Mg ratio	Ratio	3–9
Available iron (Fe)	mg/kg	100–400
Available manganese (Mn)	mg/kg	25–100
Available zinc (Zn)	mg/kg	5–30
Available copper (Cu)	mg/kg	1–15
Available boron (B)	mg/kg	0.5–5
Available N (N as nitrate)	mg/kg	20–60

Table 6.27.   Acceptable physical properties for the aggregate component of structural soil

Test method	Physical properties	Specification
AS1141.11(2009)	AS sieve (mm)	% Passing
	63.0	100
	53	85–100
	37.5	20–65
	26.5	0–20
	19.0	0–5
	13.2	0–2
	4.75	0–1
AS1141.24 (1997)	Sodium sulphate soundness (total weight % loss)	Max. 9
	Aggregate to filler soil ratio (to AS 2758.7 – 2016)	4.5–5.5

Part C. Example components for the soil supplier

The following table outlines suggested components that may likely meet the physical requirements of this specification. This is not part of the product specification. It is an example for the edification of the soil supplier of what might meet the product specification.

Example components likely to meet the physical requirements of this specification:

Nominal 63 mm hard rock aggregate (usually basalt, diorite or granite)	1 m³
Filler soil (preferably loam to clay loam)	200 L

Example base level requirements for fertilisers to be added to the filler soil (to be verified by laboratory testing and per agronomist's report):

Filler soil clay loam, sandy clay or clay	80–90% v/v
Composted soil conditioner conforming with AS 4454	10–20% v/v
Gypsum	500 g/m³ of filler soil
Urea	500 g/m³ of filler soil
Iron sulphate	1.5 kg/m³ of filler soil
Magnesium sulphate	400 g/m³ of filler soil
Lime or dolomite	600 g/m³ of filler soil
Potassium nitrate	500 g/m³ of filler soil
Superphosphate	500 g/m³ of filler soil
Trace element mix	300 g/m³ of filler soil
8–9 month controlled release	2 kg/m³ of filler soil

For the purposes of tendering, the contractor must allow for the inclusion of the above soil amendments, but the specific amendments required must be verified by laboratory testing and agronomist's recommendations.

See **Specifications G1**, **G2** and **G3** for validation, certification and test method specifications.

SPECIFICATION F2: RAINGARDENS, BIOFILTRATION AND STORMWATER FILTRATION SOILS

Part A. 'Fit-for-purpose' performance description

This specification describes a high permeability loamy sand medium for use as the growing and filtration layer in biofiltration bed installations. The specification is based on a modified version of *Adoption Guidelines for Stormwater Biofiltration Systems (Version 2)* (CRC for Water Sensitive Cities 2015). The permeability requirements are quite strict and usually mean that naturally occurring materials will not meet the specifications and mixtures of sand with some soil are required. The CRC recommendation is that only the surface 100 mm of filtration media must be fertilised to aid plant establishment. Fertiliser is recommended as a once-off event to assist initial establishment.

The required properties of the drainage layer and transition layer (if needed) are also specified.

When considering variation (**Specification G1**), more emphasis must be placed on compacted permeability than on strict adherence to particle distribution. Particle size distribution is considered informative and critical values lie within the hydraulic conductivity of the media. In practice, hydrologists may define the permeability rate more closely than is specified here following hydraulic loading calculations.

Part B. Product specification (technical parameters)

Generally, the soil must be free of 'unwanted material' and must meet all the requirements of Tables 6.28 and 6.29. Where engineers have otherwise specified permeability that specification will over-ride permeability from Table 6.28.

A typical biofiltration system will require a filter media layer (as defined below), along with transition and drainage gravel layers. In order for each layer to be compatible, hydraulic conductivity and particle bridging must be taken into account:

- *Hydraulic conductivity*: the transition layer must have a hydraulic conductivity that is higher than that of the filter media. Similarly, drainage gravel must have a higher hydraulic conductivity when compared to the transition layer. Do not use geotextile fabrics between the gravel and the biofiltration soil or the transition layer.
- *Particle bridging*: in order to determine if the particles of overlying layers pose risk of migration into lower layers, causing blockage, the determination of D-values is required. The data from a particle size analysis of each layer may be used to calculate D-values.

The CRC *Adoption Guidelines for Stormwater Biofiltration Systems* (Payne *et al.* 2015) suggest the following calculations to determine if biofiltration layers pose a risk of particle migration:

- Filter media and transition layer bridging criteria: the smallest 15% of sand particles must bridge with the largest 15% of filter media particles (CRC for Water Sensitive Cities 2015): D15 (transition layer) ≤ 5 × D85 (filter media)
- Drainage layer bridging criteria D15: (drainage layer) ≤ 5 × D85 (transition media)

Granted the D-values of each layer fall within the confines of the above calculations, the risk of particle migration is low.

Certification and 'fit-for-purpose' statement to be submitted (with any soil amendments needed). Refer to **Specifications G1** and **G2** for the testing and certification requirements.

Table 6.28.   Physical properties

Property	Units	Target range
Texture, preferred range	n/a	Loamy sand
Permeability (ASTM)	mm/h	100–300
Particle size distribution		
2.0–3.35 mm fine gravel	% w/w	< 3
1.0–2.0 mm coarse sand	% w/w	4–10
0.25–1.0 mm medium and coarse sand	% w/w	40–60
0.15–0.25 mm fine sand	% w/w	10–30
0.05–0.15 mm very fine sand	% w/w	5–30
< 0.05 mm silt plus clay	% w/w	< 3

Table 6.29.   Chemical properties

Property	Units	Target range
pH (1:5 in water)*	pH units	5.5–7.5
Electrical conductivity (1:5)	dS/m	< 1.2
Phosphorus (Colwell)	mg/kg	< 80
Total nitrogen	mg/kg	< 1000
Organic matter	% w/w	2–5

Chemical properties of the surface layer are not subject to performance specifications by CRC.
*pH requirements for the selected species to be grown within the filter media may allow for more alkaline conditions to be considered acceptable.

Part C. Example components for the soil supplier

The following table outlines the suggested components that may likely meet the physical requirements of this specification. This is not part of the product specification. It is an example for the edification of the soil supplier of what might meet the product specification. The arrangement of soil elements is shown in Fig. 6.1.

Example suggested components for the surface layer:

Loamy sand or sandy loam soil	< 20% v/v
Medium sand	70–80% v/v
Composted soil conditioner conforming with AS 4454	10–20% v/v

Example base level requirements for fertilisers for the surface layer (to be verified by laboratory testing and per agronomist's report):

Organic fertiliser (e.g. poultry manure)	5 kg/m^3 or 500 g/m^2
Compound fertiliser (NPK 16:4:14)	0.4 kg/m^3 or 40 g/m^2
Trace element mix	0.1 g/m^3 or 10 g/m^2
Superphosphate	0.2 g/m^3 or 20 g/m^2
Magnesium sulphate	0.3 g/m^3 or 30 g/m^2
Potassium sulphate	0.2 g/m^3 or 20 g/m^2

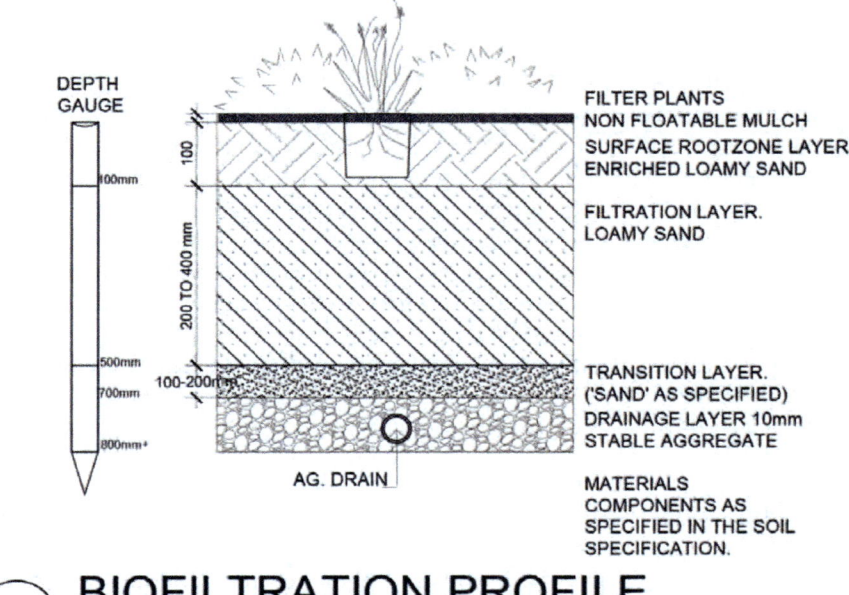

Fig. 6.1. Cross-section of the layered components described in Part C of **Specification F2**.

Example suggested components for the filtration layer:

Loamy sand or sandy loam soil	< 20% v/v
Medium sand	70–80% v/v
Composted soil conditioner conforming with AS 4454	10–20% v/v

Example suggested components for the transition layer:

Medium sand	100% v/v

Example suggested components for the drainage layer:

2–5 mm drainage gravel	100% v/v

For the purposes of tendering, the contractor must allow for the inclusion of the above soil amendments, but the specific amendments required must be verified by laboratory testing and agronomist's recommendations.

See **Specifications G1**, **G2** and **G3** for validation, certification and test method specifications.

SPECIFICATION F3: WETLAND SOILS

Part A. 'Fit-for-purpose' performance description

A sandy to silty clay soil mix designed to provide reasonable wet strength and to be stable and not dispersive. Good water-holding capacity is important to sustain plant life during extended dry periods. The optimum type of soil to achieve both good water-holding capacity and reasonable wet strength is in the fine sandy loam to fine sand texture range. Imported soils should be of a sandier nature but site soils are often clayey and can still be used provided they are not subject to traffic. To achieve the required chemical properties, a fertiliser and/or appropriate organic component is usually needed. Gypsum may be needed to prevent clay dispersion. A long-term slow-release fertiliser is recommended to allow rapid establishment of wetland plants with minimum risk of nutrient pollution of the water column.

The presence of lime or alkalinity is not considered to compromise wetland vegetation.

Part B. Product specification (technical parameters)

Generally, the soil must be free of 'unwanted material' and must meet all the requirements in Tables 6.30 and 6.31.

Table 6.30.   Physical properties and particle size analysis

Property	Units	Target range
Imported soil texture	n/a	Sandy loam to sandy clay loam
Site soils texture	n/a	Sandy loam to medium clay
Large particles	% by mass	2–20 mm = < 20% > 20 mm = < 10%
Organic matter content	% w/w	2–5
Water repellence	seconds	> 60

Table 6.31.   Chemical properties

Property	Units	Target range
pH in water (1:5)	pH units	5.4–8.0
pH in CaCl2 (1:5)	pH units	5.2–7.5
Electrical conductivity (1:5)	dS/m	< 0.5
Exchangeable Na percentage	% of ECEC	< 15
Exchangeable calcium (Ca)	% of ECEC	Normal soil 60–80 Alkaline soil 70–90
Exchangeable Ca:Mg ratio	Ratio	3–9
Available phosphorus	mg/kg	< 50
Available nitrogen (N as nitrate)	mg/kg	20–50

Part C. Example components for the soil supplier

The following table outlines suggested components that may likely meet the physical requirements of this specification. This is not part of the product specification. It is an example for the edification of the soil supplier of what might meet the product specification.

Example components (likely to meet the physical requirements of this specification):

Fine to medium grade washed sand	50–80% by volume
Sandy loam or site-won soil	10–40% by volume
On-site clay loam or clay topsoil or subsoil	10% by volume

Site soil of a clayey texture can be used provided it is not dispersive. It is easier to plant such soils before inundation to prevent pugging of clay.

Base level requirements for fertilisers (to be verified by laboratory testing and per agronomist's report):

Lime and/or dolomite	2 kg/m^3 at mixing
Compound NPK controlled-release fertiliser 9–18 month release rate	1.0 kg/m^3 or 50 g/m^2 during planting
Gypsum	300 g/m^3 at mixing

For the purposes of tendering, the contractor must allow for the inclusion of the above soil amendments, but the specific amendments required must be verified by laboratory testing and agronomist's recommendations.

See **Specifications G1**, **G2** and **G3** for validation, certification and test method specifications.

PART IV G VALIDATION SPECIFICATIONS

How to use the Part IV G specifications

There are three template specifications for the implementation phase. These are:

- G1 Quality assurance and control
- G2 Hold points and 'fit-for-purpose' statements
- G3 Compliance certification.

These three specifications are applicable to all projects. Cut and paste these specifications and fill in the appropriate specification reference number (e.g. C1 and D2) where indicated in *<bold/italic>*.

Note: make suitable additions to the implementation phase template specifications to complete your project.

PDF templates of the soil specifications are available at <https://www.publish.csiro.au/book/8226>.

Note: these specifications are typical only, not site specific, and may require specialist advice where site conditions or performance criteria is outside the parameters of these typical specifications.

These typical specifications have been provided for use by proficient and experienced landscape design professionals who are able to determine when these specifications are suitable. Where there is any doubt, consult with appropriate experienced professionals such as restoration ecologists, soil scientists or experienced landscape architects/consulting arborists.

Next step: Part IV

Add the applicable Landscape Construction Specifications (Part IV) into the specification documentation.

_____ **Specification G1** Quality assurance and control.

_____ **Specification G2** Hold points.

_____ **Specification G3** Compliance certification 'fit-for-purpose' statement and test method references.

The following typical specifications contain the elements required to ensure adequate specification at construction to ensure the specifications are met. These specifications can be used as a template and guide for site specific projects.

SPECIFICATION G1: QUALITY ASSURANCE AND CONTROL

Part A. General description

The contractor must use analytical testing to verify compliance with the product specification. This is done in two parts: initial compliance certification and quality control, as described below.

Initial compliance certification

Before any soil installation, the contractor or soil manufacturer will submit samples of trial blends likely to meet the performance specifications to a testing laboratory. See Part C of each specification for suggested formulations to start this process. The trial blend must be based on available test information on components and, if necessary, employ an agronomist for advice.

Submit trial samples to the testing laboratory, allowing sufficient time for testing and re-formulation in the case of failure to satisfy the performance criteria. Once compliant, a test certificate clearly stating compliance with the applicable criteria must be presented to the site supervisor or quality officer.

Note that alternative test methodologies may be accepted and certified as compliant by an independent expert agronomist or soil scientist.

Non-compliance will automatically generate *hold point 1.* No soil will be installed until initial compliance certification has been demonstrated.

Manufacturer's product representation: For imported soils from manufacturers, a 'product representation' document produced by the supplier may be accepted as a compliance certificate if:

- it is an off-the-shelf product line, not a custom mix
- a representative test certificate is available and is acceptably recent (within 6 months)
- the testing covers all those criteria in the performance specification
- the manufacturer's quality assurance system is externally certified.

Record keeping

Growing media initial compliance certification records must be kept in an easily retrievable manner that provides for traceability of purchase and location on site. Each compliance certification for all the product specifications used on site must be identified by date, quantity to be supplied and a copy of the formulation used to reach compliance.

Quality control: compliance during construction

The contractor must submit samples of blended soils or imported soil mixes at regular intervals during construction for the purposes of demonstrating continued compliance as part of quality control.

Test submissions

Submit representative samples of ~5 kg of each product specification, packed and labelled to indicate the source and the specification to be met. The samples must be taken in a representative manner.

The contractor must refer to the testing frequencies indicated in Table 6.32. Variations to the frequencies in this table are permitted on the submission to the superintendent of an alternative testing program that clearly achieves the desired outcome of quality control. Materials supplied from operations that have a third-party-endorsed quality assurance program may be acceptable pending submission of the relevant documentation.

Table 6.32.   Outline of the required testing frequency to achieve compliance testing. Samples must be tested to the performance criteria indicated in the product specification.

Specification	Activity	Minimum quality control test frequency
B2	Subgrade and subsoil preparation	1 per 500 m^3 or 1 per 2000 m^2 for *in situ* soils following amelioration
B3	Imported subsoils	1 per 500 m^3
C1, C2, C3	Turf and lawn soils	1 per 1000 m^3
D1, D2, D3, D4, F3	Mass planting and garden soil specifications, wetland soils	1 per 500 m^3
E1, E2, E3, E4, F1, F2	Artificial containerised growing media, structural support soils, raingardens and stormwater filtration soils	1 per 100 m^3

Note: where the delivery is less than the stated quality control testing frequency, the initial compliance certification certificate must be deemed to demonstrate compliance.

Testing

All testing as required by the product specifications must be arranged by the contractor, and carried out by the principal's nominated soil-testing laboratory. All test results records will be made available to the superintendent or quality officer.

Hold point 1

The test certificate will be accompanied by a statement of compliance from a competent person (e.g. qualified agronomist, horticulturist or soil scientist).

Compliance certificates will be in the form of a letter or short report clearly stating the material is compliant, with an attachment showing the test result relied upon. In the case of minor non-compliance or substantial compliance, a clear statement must be obtained from a qualified independent agronomist waiving the compliance and certifying the sample is compliant with or without conditions.

Non-compliance

In the case of substantive non-compliance, Hold points 2 and 3 will occur (see **Specification G2**) – one to correct soil already installed and another to ensure new deliveries are compliant.

In the case of minor non-compliance or substantial compliance, a clear statement must be obtained from a qualified independent agronomist waiving the compliance and certifying the sample is fit for purpose.

Non-compliance with the target range criteria does not necessarily render a soil not fit for purpose, but making this judgment requires an expert person to take responsibility for such deviation. Also, a conditional compliance certificate may be issued requiring that a certain fertiliser or further organic matter or some other amendment be added, with the aim of achieving compliance.

Where the drainage layer is coarser than around 5 mm, a transition layer may be needed between it and the filtration soil media to prevent soil migrating into the drainage gravel layer. Generally this will be an intermediate very coarse sand or fine gravel. Do not use geotextile fabrics over the drainage layer to prevent soil migration.

Hold point 2

The contractor will need to make corrective procedures to bring any soil that has been placed within substantial compliance in accordance with any agronomist's advice.

Hold point 3

In the event that quality control samples show substantial non-compliance from the approved performance requirements, the supplier must demonstrate compliance of any future loads. This may require re-formulation or alteration to existing formulations and may require the advice of a qualified person to meet correct analysis, and make adjustments to mixing ratios, additives and procedures to achieve compliance.

Record keeping

Growing media construction and quality control compliance records must be kept in an easily retrievable manner that provides for traceability of purchase and location on site. Each batch of soil must be identified by date of manufacture, quantity and a corresponding test result, and must link into when the material was delivered and where the material was placed.

SPECIFICATION G2: HOLD POINTS AND 'FIT-FOR-PURPOSE' STATEMENTS

Insert **Specification G2** into your landscape specification and fill in the appropriate specification reference number (e.g. C1 and D2) where indicated in **<bold/italic>**.

There are three applicable hold points:

- **Hold point 1:** initial compliance certification before installation
- **Hold point 2:** corrective procedures and their certification in the case of non-compliant produce being installed
- **Hold point 3:** re-certification in the case of on-going non-compliance.

Usually, all three of the hold points will be selected for all projects.

Hold point 1

	Completion of <***insert specification***> initial compliance certification
Process held:	Placement of <***insert specification***> soil
Acceptance criteria:	Demonstrated compliance with the <***insert specification***> Soil Specification
Release of hold point:	Submission of laboratory test certificates to superintendent together with supplier's, contractor's or independent agronomist's report certifying compliance including acceptance of any non-compliance with or without conditions

Hold point 2

	Compliance failure of <***insert specification***> during ongoing compliance certification
Process held:	Further placement of <***insert specification***> soil
Acceptance criteria:	Corrective procedures specification from a qualified horticulturist, agronomist or soil scientist for soil corrective amendments likely to result in compliance with the <***insert specification***>
Release of hold point:	Submission of laboratory test certificates to superintendent together with an independent agronomist's report certifying the corrective procedure has resulted in compliance including acceptance of any non-compliance with or without conditions

Hold point 3

	Compliance failure of <***insert specification***> during ongoing compliance certification
Process held:	Further deliveries of <***insert specification***> soil
Acceptance criteria:	Corrective procedures specification from a qualified horticulturist, agronomist or soil scientist for corrective amendments to the formulation likely to result in correction of non-compliance with the <***insert specification***>
Release of hold point:	Submission of laboratory test certificates to superintendent together with an independent agronomist's report certifying the corrective procedure has resulted in compliance including acceptance of any non-compliance with or without conditions

An example 'fitness-for-purpose' statement can be found in **Appendix B**: Forms and templates.

SPECIFICATION G3: COMPLIANCE CERTIFICATION 'FIT-FOR-PURPOSE' STATEMENT AND TEST METHOD REFERENCES

Prior to the lifting of a hold point, a declaration of compliance (certificate of compliance) shall be issued by a competent approval authority. This may be issued by the testing laboratory analyst, an independent agronomist or soil scientist or any other appointed agent of the principal contractor. In any case the identity of the competent authority shall be clearly stated.

A declaration of compliance may be issued under one of three circumstances:

1. The tested material complies with all target criteria.
2. The tested material shows minor or insignificant non-compliances with the target criteria and is deemed by the soil scientist as 'fit-for-purpose'.
3. The tested material shows non-compliances that may be corrected, in the judgment of the competent approval authority, that will bring the test material within compliance or minor non-compliance. The competent approval authority will state the corrective steps or additives that may be required.

Where soil properties cannot be corrected in any practical way a compliance statement cannot be issued and the soil must be re-formulated.

Such compliance declarations shall be signed and dated by the competent approval authority and submitted to the project manager and landscape architect before the lifting of the hold point.

An example of a declaration of compliance follows.

Declaration of compliance

Certificate no.:
To: *<insert addressee's name>*
Of: *<insert company name>*
Phone:
Fax:
Email:

Project name:
Project location:
Product name:
Supplier:
Supplier's batch
Compliance standard:
Date sampled:
Laboratory ID
Batch no. and sample no.:

I, *<insert name>*, of *<insert company name>*, having been appointed by *<insert principal's name>*, hereby certify that:

1. I am a qualified soil scientist, agronomist or analyst.
   or
2. I am a person experienced and competent in the interpretation of soil-test results for the establishment and cultivation of plants in amenity horticulture and have been appointed by the Principal or their agents.

3. This sample has been submitted by *<insert name of person submitting sample>* and has been analysed in accordance with specification *<insert reference to specification>*.
4. The extent of sampling and the results of all tests carried out on the subject soil mix conducted for the subject project are described in my report *<insert report no.>* dated *<insert date>* and are attached to this declaration for reference.
5. In my professional opinion, the soil mix described in the attached report complies with the nominated soil specification having given due consideration to the intended use and purpose, under the following circumstance:
    *   The soil complies with all target criteria with no further amendment.
    *   The soil shows minor or insignificant deviation from some of the target criteria that do not affect its fitness for purpose and do not require corrective action.
    *   The soil shows significant deviation from one or more of the target criteria but may be considered compliant if the corrective action as stated below is taken.

Corrective actions required

This certificate is issued on the understanding that the following corrective actions will be undertaken by a competent person. These corrective actions are intended to adjust *<insert purpose of the corrective action>*.

The corrective actions required are:

<Insert list of corrective actions>

This professional opinion is furnished to *<insert addressee's name>* as a representative of *<insert company name>* for their purposes alone on the express condition that it will not be relied upon by any other person and does not remove the necessity for the normal inspection of site conditions, workmanship and product liability at the time of construction.

Signed: ....

Date of report:

Test method references

Tables 6.33 to 6.34 outline the test methods referenced in the Product Specifications. These tables and the specifications G1 to G3 must form part of every landscape soil specification package to ensure full compliance.

The test methods are generally suitable and accepted for Australia and New Zealand, and the same or similar test methods are used overseas; however, expect that target ranges and some variance will occur across the world and modify with a local experienced soil scientist in the project site's region. Refer to **Appendix A3–A6** for further descriptions and comments on how these or alternative test methods can be used overseas or interpreted.

Table 6.33. Chemical tests for landscape soils

Test method name	SSSA Part 3 Chapter Reference	Rayment and Lyons Reference
pH 1:5 H_2O and $CaCl_2$	16	4B3
Electrical conductivity (EC 1:5)	14	3A1
pH in water (1:1.5)		AS 3743 Appendix D
Electrical conductivity (1:1.5)		AS 3743 Appendix D

Test method name	SSSA Part 3 Chapter Reference	Rayment and Lyons Reference
Chloride (1:1.5)		AS 3743 Appendix D
Ammonium-N (NH$_4$ 1:1.5)		AS 3743 Appendix D
Ammonium-N + nitrate-N (NH$_4$ + NO$_3$ 1:1.5)		AS 3743 Appendix D
Organic matter content	34	6B1 and 6B3
Exchangeable cations Na, K, Ca, Mg	40	15A1 in alkaline soils and 18F1 in acid soils
Available nutrients K, Ca, Mg, P, Fe, Mn, Zn, Cu, B		18F1
Available Phosphorus Acid soils method	32	18F1
Alkaline soils method	32	9B1 or 9C1
Phosphorus (Colwell)		9B
N as Nitrate		7B, 7B1 & 7B2
Ammonium (NH$_4$$^+$)		7C2a or b
Nitrogen Drawdown Index		AS 3743 Appendix E

Table 6.34. Physical testing for landscape soils

Test method name	SSSA Part 4 Chapter Reference	Other references
Particle size analysis	2.4.3	ASTM International (2010) F1632–03,
AS sieve		AS1141.11(2009)
Sodium sulphate soundness		AS1141.24 (1997)
Water repellence		AS 4419 Appendix C
Permeability (ASTM)		ASTM F1815-97
Permeability (Mc&J)		McIntyre and Jacobsen
Estimated Permeability		Hazelton and Murphy (2016) Table 2.7
Texture		AS 4419 Appendix K
Rock or foreign materials		Air dry to 40 Deg C and separate over 50 mm sieve.
Air filled porosity (AFP)		AS 3743 Appendix B
Water Holding Capacity (WHC)		AS 3743 Appendix B
Saturated repacked density		AS 3743 Appendix B by calculation
Toxicity Index		AS3743 Appendix F
Aggregate to Filler Soil Ratio		Grabosky *et al.* (2000)

* Note – refer to **Appendix A6** discussing several industry hydraulic conductivity and permeability test methods and sampling pitfalls, where misleading results can arise, particularly on highly disturbed samples. It is not possible to obtain accurate permeability measurements on aggregated clay soils by repacking them into cylinders, so an estimate method using texture and structure is used. The specifications of physical properties must be interpreted carefully by skilled and experienced soil professionals.

Bibliography – test methods

ASTM International (2010) *F1632–03, Standard Test Method for Particle Size Analysis and Sand Shape Grading of Golf Course Putting Green and Sports Field Rootzone Mixes.* ASTM International, West Conshohocken, PA.

ASTM International (2018) *F1815–11, Standard Test Methods for Saturated Hydraulic Conductivity, Water Retention, Porosity, Particle Density, and Bulk Density of Putting Green and Athletic Rootzones.* ASTM International, West Conshohocken, PA.

Bannerman SM, Hazelton PA (1990) 'Soil landscapes of the Penrith 1: 100 000 sheet'. Soil Conservation Service of NSW, Sydney.

Bassuk N, Denig BR, Haffner T, Grabosky J, Trowbridge P (2015) *CU-Structural Soil®: A Comprehensive Guide.* Urban Horticulture Institute, Cornell University.

Bodman K, Sharman KV (1993) *Container Media Management.* Queensland Department of Primary Industries, Brisbane.

Charman PEV, Murphy BW (Eds) (1991) *Soils: Their Properties and Management.* Soil Conservation Service of NSW, Sydney University Press, Sydney.

Craul PJ (1985) A description of urban soils and their desired characteristics. *Journal of Arboriculture* 11(11), 330–339. doi:10.48044/jauf.1985.071

Craul PJ (1992) *Urban Soil in Landscape Design.* John Wiley and Sons, New York, NY.

Craul PJ (1999) *Urban Soils Applications and Practices.* John Wiley and Sons, New York, NY.

Craul TA, Craul PJ (2006) *Soil Design Protocols for Landscape Architects and Contractors.* John Wiley and Sons, New York, NY.

CRC for Water Sensitive Cities (2015) Appendix C: Guidelines for filter media in stormwater biofiltration systems. Version 4.01. CRC for Water Sensitive Cities, Clayton.

Department of Mineral Resources (1983) 'Geological Survey of NSW. Sydney 1: 100 000'. Geological Series Sheet 9130. Department of Mineral Resources, Sydney.

FAWB (2009) *Guidelines for Filter Media in Biofiltration Systems Version 3.01.* Faculty for Advancing Water Biofiltration, Melbourne.

Fertilizer Industry Federation of Australia (2006) *Australian Soil Fertility Manual.* CSIRO Publishing, Melbourne.

Grabosky J, Bassuk NL (1995) A new urban tree soil to safely increase rooting volumes under sidewalks. *Journal of Arboriculture* 21(4), 187–201.

Grabosky J, Bassuk N, Trowbridge P (2000) *Structural Soil: A New Medium to Allow Urban Trees to Grow in Pavement.* Landscape Architecture Technical Information Series. American Society of Landscape Architects, Washington, DC.

Handreck KH, Black ND (2010) *Growing Media for Ornamental Plants and Turf.* 4th edn. UNSW Press, Sydney.

Hazelton P, Murphy BW (2011) *Understanding Soils in Urban Environments.* CSIRO Publishing, Melbourne.

Hazelton P, Murphy BW (2016) *Interpreting Soil Test Results: What Do All the Numbers Mean?* 3rd edn. CSIRO Publishing, Melbourne.

Isbell RF (1996) *The Australian Soil Classification*. CSIRO Publishing, Melbourne.

Keith D (2006) *Ocean Shores to Desert Dunes: The Native Vegetation of New South Wales and the ACT*. Department of the Environment and Education, Hurstville.

Landcom (2004) *Managing Urban Stormwater: Soils and Construction: The Hip Pocket Handbook*. New South Wales Government, Sydney.

McDonald RC, Isbell RF, Speight JG, Walker J, Hopkins MS (1984) *Australian Soil and Land Survey: Field Handbook*. Inkata Press, Melbourne.

McIntyre K, Jakobsen B (1998) *Drainage for Sportsturf and Horticulture*. Horticultural Engineering Consultancy, Kambah.

Munsell (2000) *Munsell Soil Colour Charts*. Gretag Macbeth, New York, NY.

Murphy CL (1993) 'Soil landscapes of the Gosford-Lake Macquarie 1: 100 000 sheet'. Department of Conservation and Land Management, Sydney.

Northcote KH (1979) *A Factual Key to the Recognition of Australian Soils*. 4th edn. Rellim Technical Publications, Glenside.

Payne EGI, Hatt B, Deletic A, Dobbie M, McCarthy D, *et al.* (2015) *Adoption Guidelines for Stormwater Biofiltration Systems*. Cooperative Research Centre for Water Sensitive Cities, Melbourne.

Peate N, MacDonald G, Talbot A (2006) *Grow What Where*. Blooming Books, Melbourne.

Peverill KI, Sparrow LA, Reuter DJ (1999) *Soil Analysis: An Interpretation Manual*. CSIRO Publishing, Melbourne.

Rayment GE, Lyons D (2011) *Soil Chemical Methods – Australasia*. CSIRO Publishing, Melbourne.

Salter PJ, Williams JB (1969) The influence of texture on the moisture characteristics of soils. V. Relationships between particle size composition and moisture contents at the upper and lower limits of available water. *Journal of Soil Science* **20**, 126–131. doi:10.1111/j.1365-2389.1969.tb01561.x

Smith K, May P, White R (2009) Root growth of *Corymbia maculata* in a constructed soil: the effect of profile design and organic amendment. In *The Landscape Below Ground III, Proceedings of a Third International Workshop on Tree Root Development in Urban Soils*, October 2008, The Morton Arboretum, Lisle, Illinois, USA. (Eds G Watson, L Costello, B Scharenbroch, E Gillman) pp. 13–18. International Society of Arboriculture, Champaign, IL, USA.

Solfjeld I (2009) Root growth after transplanting: the role of transplant timing, root-zone temperature, and adequate soil volume. In *The Landscape Below Ground III, Proceedings of a Third International Workshop on Tree Root Development in Urban Soils*, October 2008, The Morton Arboretum, Lisle, Illinois, USA. (Eds G Watson, L Costello, B Scharenbroch, E Gillman) pp. 230–236. International Society of Arboriculture, Illinois.

Sparks DL (Ed.) (1996) Methods of Soil Analysis. Part 3: Chemical Methods. Number 5 in the Soil Science Society of America Book Series. Soil Science Society of America, Inc., American Society of Agronomy, Inc. Madison, Wisconsin, USA.

Sparks DL (Ed.) (1996) Methods of Soil Analysis. Part 4: Physical Methods. Number 5 in the Soil Science Society of America Book Series. Soil Science Society of America, Inc., American Society of Agronomy, Inc. Madison, Wisconsin, USA.

Stace HCT, Hubble GD, Brewer R, Northcote KH, Sleeman JR, *et al.* (1972) *A Handbook of Australian Soils*. Rellim Technical Publications, Glenside.

Standards Australia (1997) *AS 1141.24-1997, Methods for Sampling and Testing Aggregates – Aggregate Soundness – Evaluation by Exposure to Sodium Sulfate Solution*. Standards Australia International, Sydney.

Standards Australia (1999) *AS 1289, Testing Soils for Engineering Purposes*. Standards Australia International, Sydney.

Standards Australia (2003a) *AS 3743-2003, Potting Mixes*. Standards Australia International, Sydney.

Standards Australia (2003b) *AS 4419-2018, Soils for Landscaping and Garden Use*. Standards Australia International, Sydney.

Standards Australia (2009) *AS 1141.11.1-2009, Methods for Sampling and Testing Aggregates – Particle Size Distribution – Sieving Method*. Standards Australia International, Sydney.

Standards Australia (2012) *AS 4454-2012, Composts, Soil Conditioners and Mulches*. Standards Australia International, Sydney.

Stewart A (2012) *Creating an Australian Garden*. Allen and Unwin, Crows Nest.

Thomas A, Lauricella J (2009) *Horticultural Soils 1 and 2*. Northern Sydney Institute of TAFE, North Sydney.

USGA (1993) Standard method of particle size analysis and grading sand shape for golf course putting green root zone mixes. *United States Golf Association Green Section Record* **31**(2), 28–29.

Urban J (2008) *Up By Roots: Healthy Soils and Trees in the Built Environment*. International Society of Arboriculture, Champaign, IL.

Young A, Young R (2001) *Soils in the Australian Landscape*. Oxford University Press, Melbourne.

Appendix A: Technical information, sampling and test methods

A1 TAKING *IN SITU* SOIL SAMPLES

Sampling smaller projects

It is within the abilities of a landscape architecture team or landscape contractors to take samples of topsoil from small projects and from stockpiled site soil or delivered soil for quality assurance purposes.

Composite sampling of in situ soils

Samples of topsoil representative of an area of land can be taken from individual bore holes or, in order to represent the 'average' condition and reduce sample numbers, samples may be 'composited', meaning that the sample has been blended from several subsamples taken across the area. This may be the size of a yard or a single garden bed, lawn or other uniform area.

1. Subsamples should be taken using a spade and taking a 'slice' of soil at the same depth (usually 0–200 mm), and using the same method, and then blending using the same amount, for example a cup-full of each thoroughly mixed together in a bucket to get a representative blended or composited sample for testing. For chemical testing usually 0.5–1 kg is enough but for physical testing take ~2–3 kg.
2. For topsoils 6–12 composites should be taken. Subsoils tend to be more uniform and consistent than topsoils and often a single sample from one location in a uniform soil type is sufficient. Take samples at intervals down the profile to provide information on horizon changes.
3. Blend the subsamples together ensuring they are thoroughly mixed and then remove a quantity appropriate for testing (for chemical testing usually 0.5–1 kg is enough but for physical testing take ~2–3 kg).
4. Place the sample in a labelled plastic bag, seal and record the sample location, date and who took the sample.
5. Arrange for delivery to your testing laboratory with clear instructions on what tests are required.

Sampling large and complex sites

Large complex sites either greenfield (natural condition) or brownfield (degraded disturbed industrial sites) is a skilled process requiring trained and experienced soil science skills. Designers are encouraged to employ specialist urban soil scientists as part of the design team and commission soil reuse and recovery reports for incorporation into construction specifications.

A2 STOCKPILE SAMPLING PROCEDURES

Sampling smaller (< 100 m³) stockpiles

Sampling stockpiled recovered soils on smaller projects as well as quality control testing of delivered soil stockpiles can be done by the landscape designer or contractor. For large complex projects obtaining representative samples is a skilled and trained process and the design/construct team is advised to employ the services of trained field sampling contractors.

Sampling for small projects requires a standard method of sampling that results in an average sample representative of the pile. That standard method may be varied to account for observations made at sampling (e.g. a single stockpile appears quite different at one end from the rest of the pile) and any such variation should be documented.

The following standard method provides a basis for the sampling of stockpiles. Any variation from this should be documented clearly stating the reasons. Note the method is not acceptable for sampling for contamination.

A2.1 Stockpile procedure (piles up to 100 m³)

1. Remove and discard the surface material to a depth of 100 mm. If necessary, use a shield to prevent loose particles from moving into the sampling area. Samples are best collected from ~1 m up vertically from the base.
2. Remove sufficient material, using a scoop or shovel, from each location to constitute a subsample. Take 6–12 subsamples at different places around the stockpile and of uniform size.
3. Blend the subsamples together ensuring they are thoroughly mixed and then remove a quantity appropriate for testing (for chemical testing usually 0.5–1 kg is enough but for physical take ~2–3 kg).
4. Place the sample in a labelled plastic bag, seal and record -sample location, date and who took the sample.
5. Arrange for delivery to your testing laboratory with clear instructions on what tests are required.

A2.2 Sample guidelines

Further information on sampling loads of delivered soil can be found in AS 4419–2018. The following is an extract from 'Guidelines on the taking of soil samples (informative)' from AS 4419–2018.

Soil testing should be collected with great care and an adequate number of samples should be obtained to avoid significant errors. The difficulties of defining a soil sampling procedure are well recognised and the exact procedure to be followed will be dictated by the circumstances. The following procedure is given for guidance:

a) From each load of soil to be tested, collect at least 10 representative samples, each having a volume of not less than 200 mL.
b) Without undue crushing, blend these samples to prepare a composite sample of not less than 2 L.
c) Carry out the various determinations required by this standard on the composite sample.

Note: AS 4454-2003 provides a simpler and more accessible sampling process for small stockpiles. AS 4454-2018 provides methods for professional soil contractors to sample large and complex stockpiles.

Sampling large and complex stockpiles

Sampling large complex sites with either stripped and stockpiled greenfield soils, or complex stockpiles of widely differing site-won or delivered soils, is a skilled process requiring trained and experienced sampling and/or soil science training. Designers are encouraged to employ specialist urban soil scientists or field sampling as part of the design team and commission soil reuse and recovery reports for incorporation into construction specifications.

A3 SOIL TESTING METHODS

Soil testing methods vary greatly from country to country and from state to state and laboratory to laboratory. There are regional preferences and agronomists and soil scientists will disagree about the most appropriate methods. This is most often influenced by their particular experience – the more often you have used and interpreted a test, the more confident you become applying it to recommendations.

It should always be kept in mind that conventional soil testing has its origins in agriculture and is empirically based. That means that for a given test (e.g. the tests for 'available' phosphorus), the 'normal ranges' have been established by conducting empirical experiments, nearly always on crop plants, with the establishment of a 'fertiliser response curve'. The experiment would show, for example, that, using the 'Olsen P' methods, there is no commercially viable response to added fertiliser phosphorus over 25 mg/kg. Another method, such as Bray, Mehlich, Lactic acid or Colwell, would provide a different response level. *None of these methods has been calibrated for urban landscape plants* and we are left making purely judgemental decisions. With experience, however, such decisions, if made conservatively, can provide useful advice when in the hands of experienced people.

The SESL laboratory has been offering analysis in urban landscape industry since 1984 and has developed packages of tests that are the most useful for diagnosing the widest range of soil problems. The methods are based on those published in Rayment and Lyons (2011), which is the authoritative text for Australia and New Zealand. The same or similar tests will be used overseas as well. It provides detailed methodologies for the range of tests used commonly in Australia. It does not provide any interpretation guidance. For interpretation the best references would be Handreck and Black (2010), Hazelton and Murphy (2011, 2016), Peverill *et al.* (1999) and Fertilizer Industry Federation of Australia (2006).

A4 CHEMICAL TESTS FOR LANDSCAPE SOILS

Soil test methods are given in **Specification G3**. The 'target ranges' we used in these specifications are based on the methods we refer to in the text of the specification. Chemical tests for landscape soils are given in Table 6.33 and physical tests are given in Table 6.34.

A5 PHYSICAL SOIL-TESTING METHODS

Several physical tests are required to characterise the physical properties, density, aeration and permeability properties of soils. This is critical for highly trafficked turf soils where constant compaction leads to reduction of pore space and permeability if physical properties are not carefully chosen. Unfortunately, there is no one authoritative text such as Rayment and Lyons (2011) for soil physical tests.

Particle size distribution testing uses fairly standard methodologies (sieves and sedimentation of silt and clay) and differ only in the size ranges specified. The American ASTM particle size distribution gives better resolution of sand size fractions for assessing sports field soil than the Australian Standard AS1141.11(2009).

- Soil colour employs the charts of Munsell (2000).
- For the measurement of permeability of soil the method of McIntyre and Jakobsen (1998) with use of 50 mm cylinders with no screening is considered superior to that of AS 4419-2018. Where texture is finer than heavy clay loam, use 'estimated permeability' instead of the direct hydraulic conductivity method due to matrix interference.

A6 NOTE ON HYDRAULIC CONDUCTIVITY OR PERMEABILITY TESTING

Testing of hydraulic conductivity is done on either intact or disturbed samples. Working on intact samples involves taking measurements in the field using 'infiltrometers' of various kinds or taking core samples by driving in steel cylinders and taking them back to the laboratory for testing. Testing of landscape soils for certification to a specification will be conducted exclusively using disturbed samples that are 'repacked' in the laboratory in some manner that mimics how the soil will be consolidated into place in the garden or landscape installation. We have found the McIntyre and Jakobsen (1998) methodology using 50 mm cylinders with no screening to be widely applicable to landscape soils and much superior to the AS 4419 method because it consolidates the samples to increasing degrees, providing information on how the soil will behave with increasing degrees of compaction.

All permeability methods suffer from various degrees of error and anomaly and, for this reason, need to be interpreted very carefully. Only very sandy unstructured soils show any reasonable degree of reproducibility in conductivity testing. Working with disturbed structured soils (i.e. those that are aggregated into peds) presents some profound difficulties. When such soils are repacked into cores to do permeability measurements, they can behave like gravel – thus a soil with high clay content can appear to be highly permeable, but this does not reflect its field behaviour. A common problem is that a landscape soil may be composed of a structured loam mixed with a sand. Because the soil is present as discrete peds, the overall permeability appears to be that of the sand. The soil may thus pass the particle size specification, but fail the hydraulic conductivity test because it is too highly permeable. If you increase the soil content so that it passes the permeability test, it may then fail the particle size test because the silt and clay content is too high. Where texture is finer than heavy clay loam, use 'estimated permeability' (after Hazelton and Murphy 2016) instead of the direct hydraulic conductivity method due to matrix interference.

Variations from the specifications will inevitably be required when soils don't fit the orthodox mould and such conundrums occur. The specifications of physical properties we have provided must be interpreted carefully by skilled and experienced soil professionals. The overall consideration should always be 'is it fit for purpose?' and a common sense approach should prevail over strict interpretation of permeability results that are either difficult to reproduce or just plain misleading, as with permeability measurements in disturbed structured clays. In our laboratory, we will refuse to conduct permeability measurements on strongly structured disturbed samples.

Appendix B: Forms and templates

B1 SITE SOIL ANALYSIS CHECKLIST

The following guide is a checklist for what information a specialist soil scientist or consultant needs to provide to the more complex project design process. This analysis will inform which Soil Approach Method is most appropriate to use (e.g. stockpile and condition the existing site soil or import soil). It will also provide information to inform the design, whether special solutions are required (e.g. subsoil drainage), which plants to specify and, ultimately, which soil performance specifications to select.

Table B1.   Site soil analysis checklist

Topography: • geology • slope • slope position • aspect.	Geology and topographic position are the most important determinants of soil type and will give clues to the soil type and any site constraints. Topographic position correlates with soil depth and colour. Subsoil is redder with better drainage and more yellow or even olive colours with poorer drainage. Typically, subsoils will be red at the top of hills and yellow in footslopes.
Drainage	Note the topographic position and local surface contouring. Vegetation is a good clue to waterlogging and nutrient levels (e.g. *Juncus* grasses and other sedges are typical of waterlogged conditions).
Site disturbance	Check for any unnatural contours, terracing, cuts, fill material, exposed rock/rubble or other clues of disturbance.
Vegetation type: • species present • indicator species • condition.	The type and condition of vegetation can provide information about soil and site condition (e.g. stunting, elongation, wind affected, exposed roots). The presence of introduced or endemic, invasive weeds indicates disturbance. Indicator species can point to soil fertility. The appearance of symptoms in existing vegetation provides clues into nutrient balance and availability or toxicities (e.g. reddish-purple colours in leaves and stunting indicate phosphorus deficiency). Deep green lush growth indicates high fertility. Note the presence of any plant and soil-borne pest/disease/fungi (e.g. *Armillaria*, wood rot, diseased plant material).
Soil surface condition: • compaction • salinity/salts.	Surface conditions are important indicators of soil disturbance and vegetative health. Note if the surface is vegetated, if a leaf litter (O horizon) is present and, if so, note the depth, if it is bare, crusted, if there is any surface cracking. To investigate: push a screwdriver or spade into the soil to test for compaction/resistance. Note evidence of areas that have been used as roads, hard stand (paved or gravel areas) or exposed roots. Note indications of hydrophobia (water repellence). Where salinity is present, it is often seen at the surface. Note the presence of any white salt efflorescence
Topsoil	Note: • depth • colour, use Munsell Soil Colour Chart (Munsell 2000), or use plain English such as 'yellowish-brown', 'greyish-brown' • texture, structure (Northcote 1979) • moisture condition • presence of inclusions (anthropic objects, Fe/Mn nodules, stone) • degree of compaction • presence and depth of plant roots • presence of any pallid layer (A2 horizon) on top of the subsoil • any odours (e.g. sulphidic or 'sour' odours).

Subsoil	Note:
	• depth of the boundary/interface
	• depth of subsoil
	• colour, texture, structure
	• colour and texture changes to the deep subsoil
	• depth to parent material (not always possible)
	• inclusions (e.g. ironstone, lime and gypsum accretions).

Note: this is not an exhaustive list and other factors may need to be taken into consideration for individual sites. An understanding of the functioning of soil and natural systems is needed that this checklist cannot cover. Practitioner experience and general knowledge of local site conditions also assist in site analysis.

B2 EXAMPLE SOIL ANALYSIS RESULTS

Laboratories will all differ in how they present results and very few will provide interpretation and even those that do will provide agricultural fertiliser recommendations. Normally laboratory results are sent to an agronomist who will make interpretations and recommendations for the farmer client. There are very few agronomists experienced with 'urban soil science' however.

Landscape designers are not expected to be able to interpret soil-test results so the following examples show the kind of interpretation you will need from agronomists. These example results in Fig. B2 are from SESL Australia, founded by co-author Simon Leake. SESL does make recommendations and has 40 years' experience in urban soil science and urban soil improvement for landscape projects. A range of different soil chemical properties are shown to illustrate different interpretations.

B3 SOIL TEST REQUEST FORM

Figure B3 is an example of a two-page soil-test request form developed specifically for landscape projects by SESL Australia Pty Ltd. Other test laboratories will have their own way of registering testing requests.

Soil Chemistry Profile
Mehlich 3 - Multi-nutrient Extractant

Sample Drop Off:	16 Chivers Road	**Tel:**	1300 30 40 80
	Thornleigh NSW 2120	**Fax:**	1300 64 46 89
Mailing Address:	PO Box 357	**Em:**	info@sesl.com.au
	Pennant Hills NSW 1715	**Web:**	www.sesl.com.au

Batch N°:	Sample N°:	Date Received:	Report Status: Final

Client Name:	Project Name:
	SESL Quote N°:
Client Contact:	Sample Name:
Client Order N°:	Description: **Soil**
Address:	Test Type: **FSC**

RECOMMENDATIONS

This soil sample was submitted to SESL by the client for chemical analysis. It is understood that this mix is used as a High Traffic Turf Rootzone. This soil is slightly alkaline in H_2O and very slightly acidic in $CaCl_2$ with very low salinity. Sodium and chloride are very low. The exchangeable cations indicate this soil is high in potassium. The effective cation exchange capacity (eCEC) is very low, indicating very low nutrient retention and availability. All macro and micronutrients are low. Organic matter is very low at 0.7%.

The soil chemistry of this soil is generally suited to turf underlay situations. Nutrition should be supplemented with a good all round NPK turf starter on installation.

pH is alkaline however likely suited to most turf species, for acid loving species, iron sulfate applications may be required.

Analysed by SESL Australia Pty Ltd NATA #15633

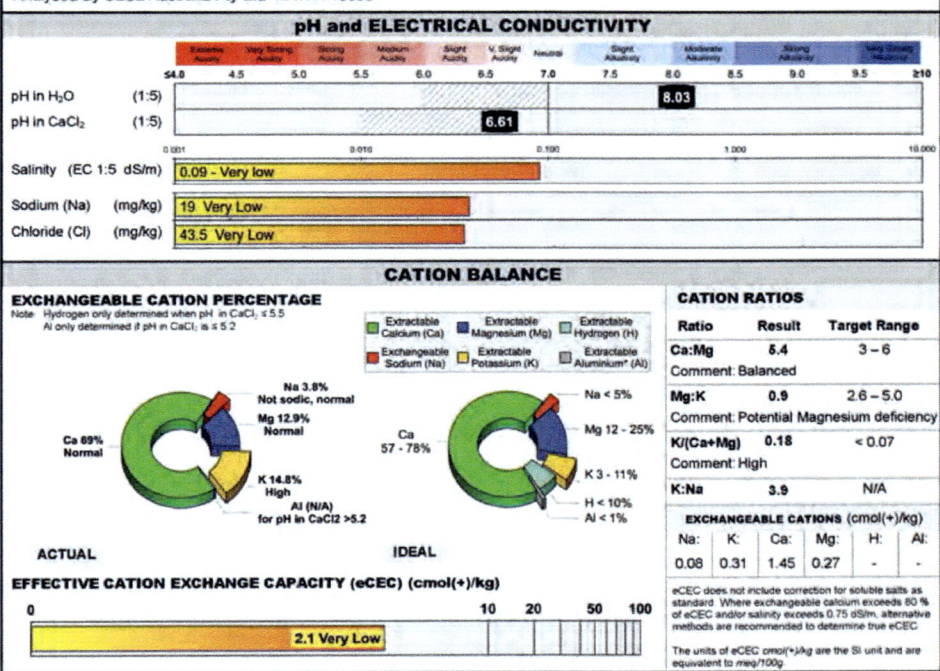

CATION RATIOS				
Ratio	Result	Target Range		
Ca:Mg	5.4	3 – 6		
Comment: Balanced				
Mg:K	0.9	2.6 – 5.0		
Comment: Potential Magnesium deficiency				
K/(Ca+Mg)	0.18	< 0.07		
Comment: High				
K:Na	3.9	N/A		

EXCHANGEABLE CATIONS (cmol(+)/kg)

Na:	K:	Ca:	Mg:	H:	Al:
0.08	0.31	1.45	0.27	-	-

eCEC does not include correction for soluble salts as standard. Where exchangeable calcium exceeds 80 % of eCEC and/or salinity exceeds 0.75 dS/m, alternative methods are recommended to determine true eCEC

The units of eCEC cmol(+)/kg are the SI unit and are equivalent to meq/100g.

A member of the Australian Soil and Plant Analysis Council (ASPAC)

Fig. B2. Soil Chemistry and Texture analysis report with recommendations. (a) Test report example 1.

Page 2

Soil Chemistry Profile
Mehlich 3 - Multi-nutrient Extractant

Sample Drop Off:	16 Chilvers Road	Tel:	1300 30 40 80
	Thornleigh NSW 2120	Fax:	1300 64 46 89
Mailing Address:	PO Box 357	Em:	info@sesl.com.au
	Pennant Hills NSW 1715	Web:	www.sesl.com.au

Batch N°:	Sample N°:	Date Received:	Report Status: Final

PLANT AVAILABLE NUTRIENTS

EFFECTIVE AMELIORATION DEPTH (mm): ● 100 ○ 150 ○ 200   **DESIRED FERTILITY CLASS:** ○ Low ● Moderate ○ High

Major Nutrients	Unit	Result	Graph	Result (g/sqm)	Desirable (g/sqm)	Adjustment (g/sqm)
Nitrate-N (NO₃)	mg N/kg	12		1.6	4	2.4
Phosphorus (P)	mg P/kg	21		2.8	8.4	5.6
Potassium (K)	mg/kg	120		16	23.7	7.7
Sulfur (S)	mg S/kg	15		2	9	7
Calcium (Ca)	mg/kg	290		38.6	168.5	129.9
Magnesium (Mg)	mg/kg	33		4.4	17.8	13.4
Iron (Fe)	mg/kg	62		8.2	73.4	65.2
Manganese (Mn)	mg/kg	6.2		0.8	5.9	5.1
Zinc (Zn)	mg/kg	1.6		0.2	0.7	0.5
Copper (Cu)	mg/kg	<0.64		0.1	0.8	0.7
Boron (B)	mg/kg	<0.1		0	0.4	0.4

Graph ranges: ☐ Very Low ☐ Low ☐ Marginal ▨ Adequate ■ High

Explanation of graph ranges:

Very Low
Growth is likely to be severely depressed and deficiency symptoms present. Large applications for soil building purposes are usually recommended. Potential response to nutrient addition is >90 %.

Low
Potential 'hidden hunger', or sub-clinical deficiency. Potential response to nutrient addition is 60 to 90 %.

Marginal
Supply of this nutrient is barely adequate for the plant, and build-up is still recommended. Potential response to nutrient addition is 30 to 60 %.

Adequate
Supply of this nutrient is adequate for the plant, and only maintenance application rates are recommended. Potential response to nutrient addition is 5 to 30 %.

High
The level is excessive and may be detrimental to plant growth (i.e. phytotoxic) and may contribute to pollution of ground and surface waters. Drawdown is recommended. Potential response to nutrient addition is <2 %.

NOTES: Adjustment recommendation calculates the elemental application to shift the soil test level to within the Adequate band, which maximises growth/yield, and economic efficiency, and minimises impact on the environment.

Drawdown: The objective nutrient management is to utilise residual soil nutrients. There is no agronomic reason to apply fertiliser when soil test levels exceed Adequate.

• g/sqm measurements are based on soil bulk density of 1.33 tonne/m³ and effective amelioration depth

Phosphorus Saturation Index

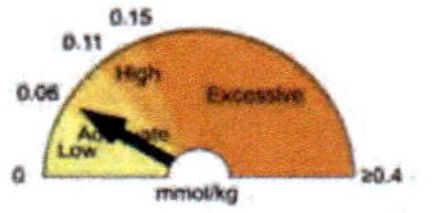

0.15
0.11
0.06   High   Excessive
0   Low   ≥0.4
0.07   mmol/kg

Adequate. Economic response to P unlikely. P application recommended maintaining current P level.

Exchangeable Acidity

Adams-Evans Buffer pH (BpH):	-
Sum of Base Cations (cmol(+)/kg):	2.1
Eff. Cation Exch. Capacity (eCEC):	2.1
Base Saturation (%):	100
Exchangeable Acidity (cmol(+)/kg):	-
Exchangeable Acidity (%):	-

Lime Application Rate (g/sqm)

– to achieve pH 6.0:	0
– to neutralise Al:	-

Calculated Gypsum Application Rate (CGAR)
(g/sqm) to achieve 67.5 % exch. Ca: 0

The CGAR is corrected for the selected effective amelioration depth (100 mm) and any Lime addition to achieve pH 6.0.

PHYSICAL DESCRIPTION

Texture:	-	Munsell Colour:	-	Organic Carbon (OC %):	Very low - 0.4
Estimated clay content:	-	Structure Size:	-	Organic Matter (OM %):	0.7
Tactually gravelly:	-	Structural Organisation:	-	Est. Field Capacity (% water):	-
Tactually organic:	-	Structural Unit:	-	Est. Permanent Wilting Point (% water):	-
Calculated EC_SE (dS/m):	-	Potential infiltration rate:	-	Est. Plant Available Water (% water):	-
Requires EC and Soil Texture result.		Est. Permeability Class (mm/hr):	-	Est. Plant Available Water (mm/m):	-
		Additional comments:			

Date Report Generated 3/01/2024

Consultant: Owen Guy

Authorised Signatory: Simon Leake

METHOD REFERENCES:
pH (1:5 H₂O) - SESL CM0002; Raymond & Lyons 4A1:2011
pH (1:5 CaCl₂) - SESL CM0002; Raymond & Lyons 4B4:2011
EC (1:5) - SESL CM0001; Raymond & Lyons 3A1:2011
Chloride - Raymond & Lyons 5A2a:2011
Nitrate - Raymond & Lyons 7D:2011
Aluminium - SESL CM0007; Raymond & Lyons 15A1:2011
P, K, S, Ca, Mg, Na, Fe, Mn, Zn, Cu, B - SESL CM0007; Raymond & Lyons 18F1:2011
Sulfer pH and Nitrogen - SESL Methods of Soil Analysis 2001, 4th 3, Ch 17, Adams-Evans (1982)
Texture/Structure/Colour - P90002 (Tjamin)
"Rainbook" (1982), Structure" - "Murphy" (1991), Colour "Munsell" (2000)

A member of the Australian Soil and Plant Analysis Council (ASPAC)

Fig. B2.   Soil Chemistry and Texture analysis report with recommendations. (b) Test report example 2.

PHONE: 1300 30 40 80
EMAIL: info@sesl.com.au

JOB CONTROL SHEET (JCS)

1 CONTACT DETAILS

COMPANY NAME		CONTACT NAME	
ADDRESS		MOBILE/ PHONE	
		EMAIL	

2 DETAILS OF PROJECT

PROJECT NAME			
PURCHASE ORDER NO.		SESL QUOTE NO.	
SESL CONSULTANT		DO YOU REQUIRE THE FOLLOWING? (PLEASE TICK)	
		RESULTS ONLY ☐ INTERPRETATIONS ☐ RECOMMENDATIONS ☐	
IS THIS FOR LEGAL PROCEEDINGS?	YES ◯ NO ◯	IF YES, CONTACT SESL FOR A CHAIN OF CUSTODY FORM (COC)	
TURNAROUND TIME (TAT)	STANDARD ◯ URGENT ◯	100% SURCHARGE FOR URGENT TURNAROUND, SUBJECT TO LABORATORY AVAILABILITY AND NATURE OF ANALYSIS.	

3 BACKGROUND TO THE REQUEST

DESCRIBE THE PURPOSE OF THE TESTING TO HELP US ENSURE THE CORRECT ANALYSIS

(E.G., MATERIAL COMPLIANCE, PLANT DISEASE, SOIL CONTAMINATION, MATERIAL CONTAMINATION, PLANT HEALTH TISSUE)

4 SAMPLES SUMBITTED

#	SAMPLE NAME	TESTING & COMMENTS
1		
2		
3		
4		
5		
6		
7		
8		
9		
10		

ADDRESS FOR SAMPLES

Attention: Sample Receipt
SESL Australia
16 Chilvers Rd, Thornleigh NSW 2120

VERSION: 2.3 (04.10.23) OFFICE USE ONLY	
DATE RECEIVED	
SAMPLE TEMP (°C)	
RECEIVED BY	
BATCH NO.	

ABN	WEBSITE	PHONE	EMAIL	LAB/POST
70 106 810 708	sesl.com.au	1300 30 40 80	info@sesl.com.au	16 Chilvers Rd Thornleigh NSW 2120

Fig. B3.   Example of a testing request sheet.

Appendix C: Soil rooting volumes for trees

C1 GENERAL OVERVIEW

While in urban situations, it is very difficult for trees to reach their optimum potential; providing appropriate volumes of suitable soil where possible will allow trees to reach an acceptable 'design response size' (design size) for their estimated lifespan. On the other hand, a perceived or actual lack of adequate soil volume based on the size of a fully mature naturally growing tree should not result in a decision to not plant a given tree species where a degree of stunting after 20 years would still result in an acceptable design outcome.

For the most part, trees will stunt in limited soil volumes due mostly to reduced water and nutrients available to them, and will have shorter useful lifespans, and yet still provide adequate function as a street tree. Climate, especially rainfall, plays a major part in deciding and designing soil volumes, as does the ability to provide supplementary irrigation in dry times.

When planted into urban settings generally tree species have more limited growth rates and lifespans than trees grown in natural, unobstructed soils. Horticultural literature generally states tree sizes in ideal growing conditions (in their natural habitat) which can often put off landscape architects from planting larger growing tree species.

When designing, consider that tree stunting is most likely in urban settings with limited soil volumes and use that stunting factor as your 'design size'. We have stated the 'design size' after 15–20 years in the Soil Volume Simulator.

A stunted tree may provide a perfectly adequate design response as intended.

Trees will be able to provide the following benefits in urban environments (generally the larger the tree the greater the benefit listed). These are just a few of the tangible benefits that large trees provide:

- increase the monetary value of urban areas and neighbourhoods, and increase business financial turnovers and retail customers
- provide shade and temperature regulation through localised evapotranspiration
- uptake excess groundwater and alleviate stormwater storage requirements
- capture air particulates (pollutants and dust) and redirect them out of the air, dissipate wind and provide visual privacy
- stabilise surrounding ground, reduce salinity and soil erosion
- support soil microbial systems and hence improve overall soil health
- provide shade which helps prolong the lifespan of footpaths, buildings and road pavements
- provide oxygen, store carbon and foster microorganisms in their root zones (critical for all life)
- promote biodiversity and vegetation corridors to support habitat links (often disconnected in urban or highly modified environments).

For further information on the benefits and the case for larger trees in urban environments, refer to the paper at <https://www.naturewithin.info/UF/TreeBenefitsUK.pdf>.

It should be noted that larger trees provide greater benefits socially, environmentally, and functionally than smaller trees. It should always be the first preference to design and install tree species and conditions that will allow trees to grow to their full potential.

The reference list at the end of this Appendix includes international references, publications on research, and tests performed to quantify optimum and minimum soil volumes. A tabulated two-part summary of over nine leading scientists, professionals and authors in the related fields is outlined in Table C5 and averages have been calculated as shown and converted into metric units, where applicable. These averages are summarised in Table C1.

The findings from these scientific and peer reviewed experiments and recommendations have been considered in the development of the Soil Volume Simulator. We have also conducted and combined tree soil-testing experiments and tree leaf analysis experiments, observations, and experiences and observations with different tree species and growing environments.

Drawing upon the eight key factors for estimating soil volumes for new trees in limited spaces, developed in the first edition of this book in 2014 (refer to Fig. C1), we have streamlined and simplified the estimating method into a simulation tool. In 2017 Elke Haege launched the online Soil Volume Simulator.

SOIL VOLUME SIMULATOR

The Soil Volume Simulator is a free to use online tool and is found at the authors' website: <https://www.elkeh.com.au/soils/>.

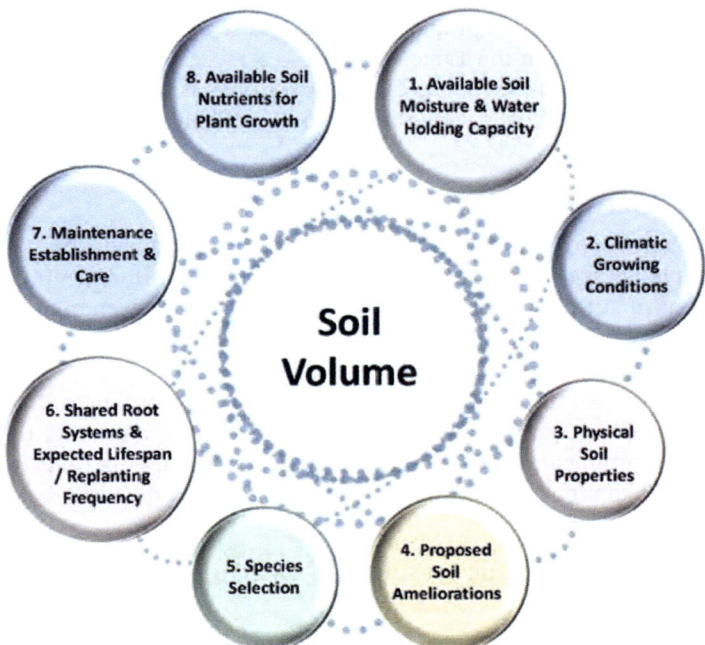

Fig. C1.   The eight key factors that need evaluation before recommending a soil volume for trees in limited spaces. The diagram shows how the key factors are interlinked in order to reach a 'soil volume' in the centre of the diagram.

Table C1. Summary of average soil volume minimum and optimum recommendations from leading industry professionals

Minimum soil volume recommendations expressed as an average and derived from published research findings as listed in table below					
Relative tree size at maturity**	Small	Medium	Large	Optimum recommendation	Refer to tables following for averages on tree sizes
Averages for research findings	13.3 m³	31.3 m³	> 18m² area*	42 m³	Measurements (converted into m³ where required)
Minimum soil volume recommendations expressed as an average and derived from regulatory documents as listed below					
Averages for published regulatory documents	11.9 m³	21.7 m³	37.9 m³	34 m³	Note: predominantly Canadian and USA sources

* This is the only reference to large trees and was expressed as a canopy area not a soil volume.
** Based on Gilman (1997), height and canopy spread from his findings.

The published research findings have been conducted on trees in Australia, the United Kingdom, Canada, the United States of America, the Netherlands and Germany, as well as other Western European countries. It should be noted that research findings have used varying species, and different growing conditions and climates to measure optimum soil growing volumes. In order to understand the complexities of the site conditions, the authors believe that, before setting a 'site specific' recommended minimum volume, eight key factors (see Fig. C1 and Soil Volume Simulator) should be considered.

Soil physical and chemical properties enable these above listed conditions at varying depths and are outlined in the Table C2.

A general rule of thumb when estimating volumes is to limit the depth to 400–600 mm as appropriate and always dig a test hole to validate your decision and account for underground services.

C2 ROOTING DEPTH

Depth plays an important role in calculating tree rooting capacities. Unless soils are almost completely free draining, with a texture class of almost full coarse sand, it is unlikely to have sufficient water-holding capacity and air exchange (roots breathe) at depths greater than 600 mm for coarse textured soil and 400 mm depth for clay loam or clay soils or soil profiles with limited drainage function.

The exception with this is where StrataCells© or similar products are utilised together with a soil profile design that provides root: air exchange at greater depths due to the reduction of slumping and compaction and able to meet the functions listed below.

Roots are able to properly function where the growing media can provide the following:

- The medium is 'loose' (i.e. not compacted).
- The medium is porous and not subject to waterlogging (i.e. aerobic conditions are present).
- Nutrients are available.
- Water is available.

Table C2.   Typical tree rooting depths

Depth category	Shallow	Medium	Deep
Soil type	Shallow podzols	Deep podzols	Sands Chernozems Kraznozems
Typical tree rooting depths (mm)	100–300	300–500	500–700

In the case of structural soils and installation of structural manufactured 'cells' such as Stratacells©, greater depths may be able to be achieved, but this will require specialist input from suitably qualified experienced industry professionals. For greater soil depths refer to Smith *et al.* (2009).

C3 SHARED ROOTING VOLUMES AND SURROUNDING SOIL

Verge zones are often adjacent to kerb and gutter footings by the roadside and footpath footings and associated structures on the other side. Verges also often contain underground services limiting tree root development.

Experiments by Solfjeld (2009) for the Norwegian Public Roads Administration demonstrate the soil volume and planting strip volume is reduced by 0.5 m^3 per metre on each side.

The example took a surface area of 10 m × 2 m (a typical verge area) × 1 m depth. By taking into account the infrastructure (footings and services) the soil volume reduces from 20 m^3 to 10 m^3.

Solfjeld outlines that increasing the verge width from 2 m to 3 m doubles the soil volume available for roots from 6 m^3 to 12 m^3 if trees are planted at 6 m intervals.

Additional methods to increase available soil volumes for trees in streets and verges include:

* providing structural manufactured cells/systems or adequate vaulted soil zones for roots, such as through cantilevered road sidings
* designing permeable adjacent areas to tree planting pits such as footpath or permeable car parking hardstand adjacent trees
* designing connected soil break-out zones
* grouping trees together in connected continuous soil rooting zones.

Appreciation of the contribution surrounding soils might make to a tree requires some investigations of the planting site and an estimate of the quality and potential effectiveness of surrounding soils. Inevitably, deciding on the contribution the surrounding soil will make, will be judgemental and will rely upon the experience of the arborist or landscape architect, and should include such observations as the performance of existing trees in the surrounding area, availability of topsoil and soil type in adjoining garden beds, and of the successful and not successful existing tree species near the project site.

C4 TREE PIT OPENINGS

The size of tree pits is to be considered for the tree to suitably anchor itself and to allow for the development of the tree's natural taper at the root crown base and to allow for the zone of root crown upheaval as the tree matures. Tree pits and tree grates also need to allow for trunk girth growth.

Shared Root Systems Diagram.

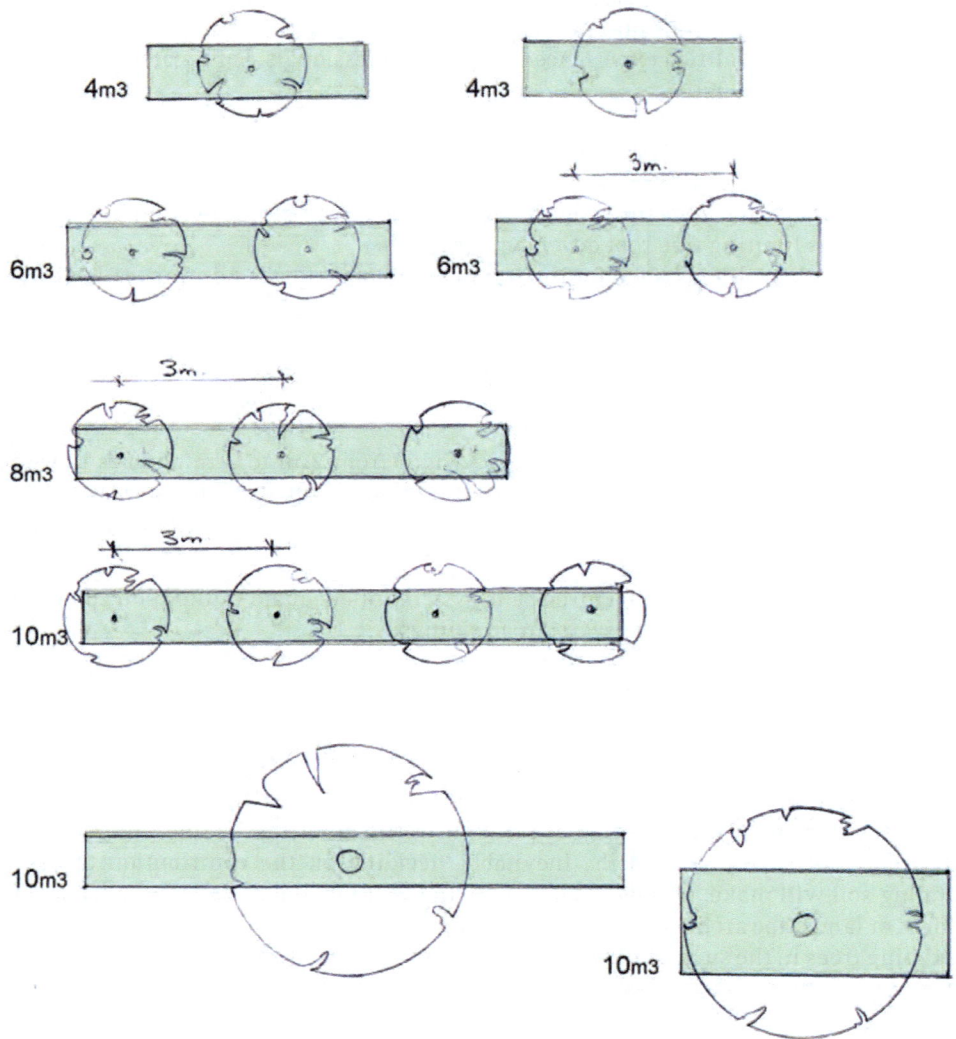

Fig. C3. Diagram illustrating the shared root system findings and considerations for designing multiple smaller growing trees versus one larger tree in similar available soil volume as shown for the 10 m^3 examples.

In highly urbanised situations, the tree pit openings may be the only opportunity for water and air exchange to the tree roots, so the bigger the tree pit opening, the better.

Some tree species form a greater trunk taper at the root crown base and some trees are more likely to put out surface roots (or even buttress roots) as they mature. It may be not be necessary to consider trunk taper if tree replacement frequency occurs before the root crown base taper develops.

The following tree pit openings can be used as a guide based on the 'design size' of the tree.

Table C3. Typical tree pit opening guidelines

Tree design size and height (at 15–20 years)	Recommended minimum tree pit opening (radius in metres from the centre of the tree)
Small to 4 m high (at 15–20 years)	0.5 m
Small/medium 4–9 m high (at 15–20 years)	1 m
Medium 7–10 m high (at 15–20 years)	1.4 m
Tall tree 9–20 m high (at 15–20 years)	1.7 m
Tall and wide: canopy ~12 m diameter and height of 8 m + and trees that develop buttress roots (e.g. *Ficus macrophylla*) (at 15–20 years)	2 m

C5 SOIL VOLUME SIMULATOR SMALLEST AND LARGEST VOLUMES FOR TREES

The seven influences (selections) in the Soil Volume Simulator are under the following headings:

1. Tree design size and height
2. Climatic growing conditions (particularly rainfall)
3. Soil suitability within the tree pit
4. Maintenance including irrigation
5. Design life or lifespan/tree replacement time (considering acceptability of tree stunting)
6. Including the surrounding soil (for determining if and how much of the surrounding adjoining soil can contribute to the total soil volume). Refer to C3 for detail on this.
7. We have included 'shared root zones' as an additional consideration to assist with planting multiple trees in the same soil zone. Refer to C3 for detail on this.

Table C4 and Fig. C4 outline the smallest and largest minimum soil volumes possible when using the Soil Volume Simulator. This does not account for influences 6 and 7 (listed above) being contributions of the surrounding soil and multiple tree planting volume modifications (refer to C3). Refer to the online Soil Volume Simulator to refine these features and tailor them to your simulated tree planting scenario.

Table C4. An outline showing the smallest and largest minimum soil volumes possible when using the Soil Volume Simulator (based on the first five categories of the Soil Volume Simulator)

Tree design size/height (at 15–20 years)	Smallest possible volume m³. Optimal conditions	Largest possible volume m³. Least optimal conditions	On slab: smallest possible volume m³. Optimal conditions	On slab: largest possible volume m³. Least optimal conditions using on slab soil type E1 + E2
Small to 4 m (at 15–20 years)	2.2	8.7	3.2	7.7
Small/medium 4–9 m high (at 15–20 years)	3.4	13.8	5.3	12.6
Medium 7–10 m high (at 15–20 years)	4.3	21.4	7.2	20
Tall tree 9–20 m high (at 15–20 years)	11.6	32.7	15.4	30.7
Tall and wide tree: canopy ~12 m diameter and height 8 m + and trees that develop buttress roots (e.g. *Ficus macrophylla*) (at 15–20 years)	17.9	43.7	22.7	41.5

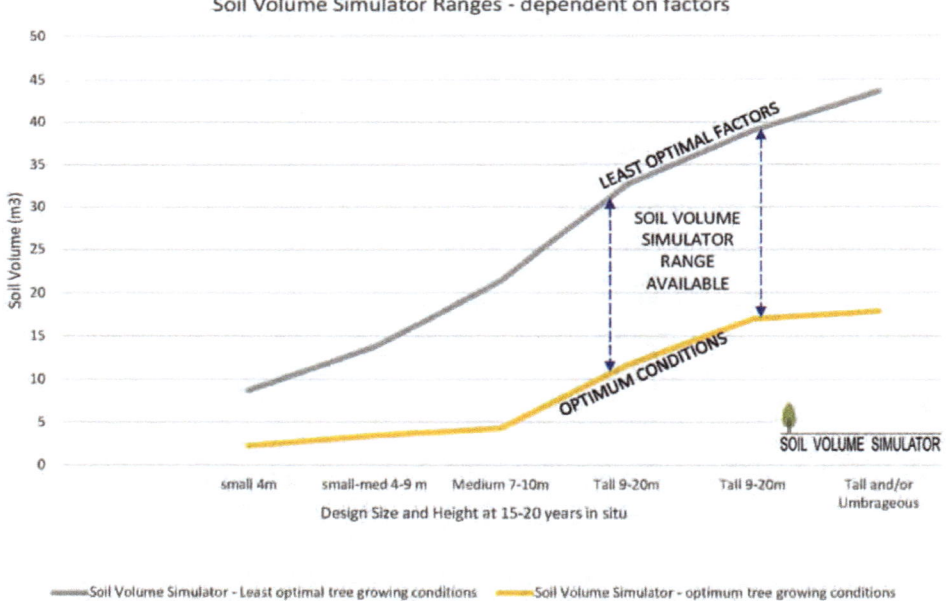

Fig. C4. Graph outlining the minimum and maximum ranges of soil volumes that can be simulated in the Soil Volume Simulator based on five key influencing factors. Note: this table does not account for factors 6 and 7 of the seven influences (sections) within The Soil Volume Simulator. The Soil Volume Simulator is a free online tool and is found at the authors' website: <https://www.elkeh.com.au/soils/>.

C6 APPLYING SCIENTIFIC RULES TO BIOLOGICAL SYSTEMS

Both professions of soil science and landscape architecture require the application of scientific rules to biological systems. This is not always straightforward, and there are always going to be exceptions to the rules, as demonstrated in Table C5 of scientific experiments at the end of this chapter.

The Soil Volume Simulator (https://www.elkeh.com.au/soils) provides typical approaches in order to simulate and estimate, but not to definitively calculate, soil volumes.

When applying and considering soil volume simulations, it is important to remember the following.

Trees rely on the surrounding soil even if their roots do not grow in the full extent of that zone.

The more appropriate and adequate the soil zone for a tree:

- the higher the value of the tree
- the lower the cost (maintenance and replacement frequency)
- the greater the efficiency and performance of the tree
- the more it contributes to the urban ecosystem, and
- the shorter time it takes for the tree to reach its potential.

The smaller the soil volume zone:

- the more susceptible the tree is to climatic stressors such as drought and flood, and nutrient deficiency
- the life expectancy of the tree is significantly reduced
- its pest and disease resilience is lowered
- it may be stunted
- it may provide poor amenity and less shade
- it is also less healthy and is more likely to be unsafe structurally and unstable.

C7 REFERENCE DOCUMENTS AND EXPERIMENTS RELATING TO RECOMMENDED MINIMUM SOIL VOLUMES

Estimating soil volumes and tree rooting volumes is not an exact science. The main purpose of the Soil Volume Simulator is to prompt the designer to consider various factors that influence rooting volume requirements for trees and to utilise this simulator to communicate design considerations to the project team and client.

The Soil Volume Simulator is a free online tool and found at the authors' website: <https://www.elkeh.com.au/soils/>.

It should not be forgotten that rainfall, and the incidence and reliability of rainfall, is probably the greatest single influence.

Table C5 outlines the industry-accepted publications and scientific experiments relating to recommended minimum soil volumes for tree rooting capacity.

Less reliable are the recommended soil volumes given by regulatory authorities, as there is often no reference provided nor explanation on how the recommended minimum soil volumes came to be. These soil volumes have therefore separately been shown in Table C6, and should not be given the priority of the scientifically evaluated volumes in Table C5.

Table C5. Industry-accepted publications and scientific findings in relation to both canopy diameters and recommended minimum soil volumes for tree rooting capacity

Canopy diameters	Small	Medium	Large	Notes source (refer to reference)
Average canopy diameters (m)	< 9	< 15	> 15	Note: measurements represent crown diameter* (canopy plane)
Planning Department of NSW (2011), stated canopy diameters (m)	4	8	16	To be used as a guide for a small, medium or large tree (at maturity in ideal conditions) Note: account for fastigiate trees that have a smaller canopy spread than umbrageous species

Recommended minimum soil volume findings	Minimum small tree soil volume (m³)	Minimum medium tree soil volume (m³)	Minimum large tree soil volume (m³)	Recommended soil volume (m³)	Notes
Urban (2008)	23			34	
Bakker (1983) cited in Kopinga (1991)	14–21	23.5–35 (for a 15 m crown diameter)			Recommends: 0.5–0.75 m³ of medium coarse sand with 5–7% organic matter per 1 m² of crown projection. Tree evaporation: 1.5 times in urban situation versus forest
Gilman (1997) (Gilman provides areas, not volumes)	Area: 4–9 m²	Area: 9–18 m²	Area: > 18 m²		Observations conducted in Florida, USA
Helliwell (1986) (average for South-east England)				44.6	
Kent et al. (2006)				42.5	
Kopinga (1991)				42.4	
Lindsey and Bassuk (1991)	18	30	40		2 ft³ per every 1 ft of crown projection
Lindsay and Bassuk (1992)	5 m³ (per tree for shared 'trenches' only and expressed as a minimum)				Research based on water requirements and evapotranspiration levels in British climatic conditions

Canopy diameters	Small	Medium	Large	Notes source (refer to reference)	
Schoenfeld (1975)				45	Netherlands, Dutch elms
Schoenfeld and van den Burg (1984)	11			45	Netherlands roadside poplars. > 10 ideal 45 m^3. 1/10th volume of rooting volume per total live canopy (in UK)
Schoenfeld and van den Burg (1984)	10			43 m^3 with 750 mm soil depth	Mature tree height in metres to correspond with the soil volume in m^3. E.g. 15 m high = 15 m^3 soil
Averages for all above entries (m^3)	13.3	31.3	Only 1 entry. Area given: > 18 m^2 (further research is needed)	42	Measurements (converted into m^3 where required)
	Small tree volume	Medium tree volume	Large tree area (after Gilman 1997)	Optimum volume	Netherlands, Dutch elms

Table C6. Outlines of industry regulatory documents and government policies relating to recommended minimum soil volumes for tree rooting capacity

Recommended minimum soil volume findings from publications/regulatory documents	Minimum small tree soil volume (m³)	Minimum medium tree soil volume (m³)	Minimum large tree soil volume (m³)	Recommended soil volume (m³)	Notes
Tree size	Small	Medium	Large	Optimum	
Planning Department of NSW (2011)	9 minimum Soil depth: 800 mm	35 minimum Soil depth: 1 m	130 minimum Soil depth: 1.3 m*		*'Large tree' defined as having a 16 m canopy diameter (volume considered to be excessive)
City of Toronto, Canada	15			30	(Advocates shared planters)
Oakville, Ontario, USA	15			30	(Advocates shared planters)
Baltimore Waterfront Harbour Initiative				42	<http://www.waterfrontpartnership.org/>
University of Florida, IFAS Extension	8.5	34	76.5		With reference to trees withstanding wind loading
West Virginia Department of Environmental Protection	14	21	27		Based on recommendations from Prince William County, Virginia, USA
Athens-Clarke County, Georgia, Guidelines, USA	5.7	12.74	22.65		
State of Minnesota Sustainable Building Guidelines (MSBG)	14.16				B3-MSBG Version 2.1, 2009
Charlotte North Carolina and Mecklenburg County	7.7				City of Charlotte Land Development Standards Manual (CLDSM)
British Colombia, Canada. Landscape Standards	6	9	20		2008, 7th edition, British Colombia Society of Landscape Architects and Nanaimo, BC, 2009 public works recommendations

Recommended minimum soil volume findings from publications/regulatory documents	Minimum small tree soil volume (m³)	Minimum medium tree soil volume (m³)	Minimum large tree soil volume (m³)	Recommended soil volume (m³)	Notes
AECOM (2009)	15	23	30	Trees in parking lots: 15 m³ min. with 1 tree every 5 car spaces. Street verges: tree pit width: 1.5 m minimum. To give 15 m³ per tree (30 m³ for large trees). 10 m maximum spacing intervals per street tree	
Aspen, Colorado and Pitkin County	7.07	28.31	63.71		
Denver Forestry Department (2011)	21.237	'... tree pit areas not accepted ... must use trenches, root zones, break out zones, structural cells, other uncompacted soil technology. 1.5 m × 1.5 m is not acceptable. '... credited soil analysis required and remediation works to be proposed'			
Emeryville precedent, California, USA	17	25.4	33.9		
City of Alexandria, Virginia, USA (2007)	8.495				'one tree per 10 car parking spaces'
'Greenleaf' Structural Soil Guidelines (n.d.)			30		Recommended equivalent without using their product
Averages from the above published regulatory documents (m³)	11.9	21.7	37.9	34	Predominantly Canadian and USA sources

The above entries have been cited from Marritz (2012). Source documents have been reviewed, verified and converted to m³ where needed.
*'Crown projection' refers to the area (circle area) out to the tree's dripline – and refers to a mature tree at its full optimum potential.
Refer to Section C8 below for full references and sources of information and studies.

C8 REFERENCE LIST AND FURTHER READING FOR TREE ROOTS AND VOLUMES

AECOM (2009) *Trees for Tomorrow: Streetscape Manual*. Markham, Ontario, Canada.

Bakker JW (1983) Groeiplaats en watervoorziening van straatbomen. *Groen* **39**(6), 205–207.

Burg J van den (1990) Minerale voeding van bomen:bladmonsteranalyse als basis voor een bemestingsadvies. *Groenkontakt* **16**(1), 10–19 (cited in Kopinga 1991).

City of Alexandria, Virginia, USA (2007) *Landscape Guidelines*, Section II, B, 2. p. 17.

Denver Forestry Department (2011) *Street Tree Plan Review Checklist*. p. 3. Denver Parks and Recreation, Denver, CO.

Gilman EF (1997) *Trees for Urban and Suburban Landscapes*. Delmar Publishers, Albany, NY.

'Greenleaf' *Structural Soil Guidelines* (n.d.) https://www.greenleafireland.com/what-does-an-urban-tree-need-to-survive/

Helliwell DR (1986) The extent of tree roots. *Arboricultural Journal* **10**, 341–347. doi:10.1080/03071375.1986.9756343

Kent D, Shultz S, Wyatt T, Halcrow D (2006) *Soil Volume and Tree Condition in Walt Disney World Parking Lots*. Board of Regents by the University of Wisconsin-Madison, Madison, WI.

Kopinga J (1991) The effect of restricted volumes of soil on the growth and development of street trees. *Journal of Arboriculture* **17**(3), 57–63. doi:10.48044/jauf.1991.016

Lindsey P, Bassuk N (1991) Specifying soil volumes to meet the water needs of mature urban street trees and trees in containers. *Journal of Arboriculture* **17**(6), 141–149. doi:10.48044/jauf.1991.040

Lindsey P, Bassuk N (1992) Redesigning the urban forest from the ground below: a new approach to specifying adequate soil volumes for street trees. *Arboricultural Journal* **16**, 25–39. doi:10.1080/03071375.1992.9746896

Marritz L (2012) *Municipalities With Soil Volume Minimums for Trees*, <https://www.deeproot.com>, accessed Oct 2012 and Jan 2013.

Planning Department of NSW (2011) *Residential Flat Design Code: Site Configuration. Deep Soil Zones. Part 02: Site Design*, p. 8. Sydney.

Schoenfeld PH (1975) De groei van Hollandse i ep in de kust provincies van Nederland. *Nederlands Bosbouw Tijdschrift* **47**, 87–95 (cited in Kopinga 1991).

Schoenfeld PH, van den Burg J (1984) Voortijdige bladval en groeiafname bij 'Heidemij'populier in beplantingen langs autowegen. *Nederlands Bosbouw Tijdschrift* **56**, 12–21 (cited in Kopinga 1991).

Smith K, May P, White R (2009) Root growth of *Corymbia maculata* in a constructed soil: the effect of profile design and organic amendment. In *The Landscape Below Ground III, Proceedings of a Third International Workshop on Tree Root Development in Urban Soils*, October 2008, The Morton Arboretum, Lisle, Illinois, USA. (Eds G Watson, L Costello, B Scharenbroch, E Gillman) pp. 13–18. International Society of Arboriculture, Champaign, IL, USA.

Solfjeld I (2009) Root growth after transplanting: the role of transplant timing, root-zone temperature, and adequate soil volume. In *The Landscape Below Ground III, Proceedings of a Third International Workshop on Tree Root Development in Urban Soils*, October 2008, The Morton Arboretum, Lisle, Illinois, USA. (Eds G Watson, L Costello, B Scharenbroch, E Gillman) pp. 230–236. International Society of Arboriculture, Champaign, IL, USA.

Urban J (2008) *Up By Roots: Healthy Soils and Trees in the Built Environment.* International Society of Arboriculture, Champaign, IL, USA.

Urban J (2009) An alternative to structural soils for urban trees and rain water management. In *The Landscape Below Ground III, Proceedings of a Third International Workshop on Tree Root Development in Urban Soils*, October 2008, The Morton Arboretum, Lisle, Illinois, USA. (Eds G Watson, L Costello, B Scharenbroch, E Gillman) pp. 301–305. International Society of Arboriculture, Champaign, IL, USA.

Vrestiak P (1987) Leaf biomass of the sycamore maple (*Acer pseudoplatanus* L.) in urban greenery. *Ekologia* **6**(1), 3–14 (cited in Kopinga 1991).

Glossary of industry terms

The majority of the following terms are defined within the context of this handbook. Other terms have been added to provide the most commonly accepted industry definition.

Acid sulphate soils (ASS) Soils, usually in coastal or estuarine areas, containing reduced sulphides. They can be potential ASS, where the sulphide remains reduced, or actual ASS, where the sulphides have been oxidised and pH has gone acidic (usually pH 4 and below). Sulphide is present as iron sulphides, primarily pyrite; hence they are black or grey in colour. When oxidised, soils turn yellow or red. Changes to the watertable are the most common cause of oxidation and acidification (*see also* sulphide).

Acidification When the pH of the soil decreases through a build-up of hydrogen cations from fertilisers, by the action of plant roots and when alkaline cations such as Ca, Mg and K are leached from the soil or taken up by plants.

Additives Products such as soil conditioners, mulches and manures which ameliorate the soil to optimise the condition through improving the physical properties such as by increasing water- and nutrient-holding capacity, and improving aeration and water infiltration..

Aeration The process of oxygenating soil by loosening and supplying air for the plant root zone.

Aerenchyma A special root structure that can develop in some plant species to allow oxygen to diffuse onto the roots. Rice, maize and barley, aquatic and wetland plants, and some plants growing in flooded landscapes can develop aerenchyma on their root systems.

Aerobic tissues Tissues that respire and need oxygen. Living plant roots contain aerobic tissues than can be killed by even short periods of hypoxia (low oxygen levels) or anaerobia (zero oxygen).

Aggregate Soil particles that are bound together more strongly than to other particles by clay and organic matter. Soil structure is determined by the type and degree of aggregation. Aggregation results in the development of pore spaces for aeration and water infiltration. Well-structured soils show optimum aeration and water-holding capacity.

Aggregate stability test (Emerson aggregate test) A test that measures the structural stability of the soil against flowing water, wind or mechanical forces. It involves placing a dry aggregate into water and observing whether it slakes and disperses.

Air-filled porosity The proportion of the total void or pore space of a soil or growing medium occupied by air at a given moisture content. In AS 3743 the air-filled porosity is that of a 12 cm high pot allowed to drain after being saturated.

Alluvium Soils deposited through erosion by rivers or flooding.

Amend or ameliorate To add any substance to improve the suitability of topsoil or subgrade, such as lime, dolomite, gypsum and compost. Addition may be via surface application, mechanical mixing or other appropriate means of ensuring satisfactory adjustment of the chemical and/or physical properties with the aim of bringing them within the specification. Amelioration also applies to any physical process used to improve rooting conditions such as compaction relief and coring. Common chemical ameliorants are lime, dolomite and gypsum.

Ammonification The breakdown of organic nitrogen to ammonium. This can occur in several ways, but primarily via bacteria, and is accelerated by alkaline conditions and warmer temperatures. Ammonification can occur through anaerobic or aerobic conditions, depending on the bacteria involved.

Anaerobic *See* anoxic.

Anoxic Completely depleted of oxygen, usually as a result of the respiration of roots and microorganisms. Common in highly organic layers waterlogged due to poor drainage. Prolonged anoxia results in sulphide production.

Apedal Single-grained, structureless soil that has no coherency (*see also* massive).

Available soil water-holding capacity (ASWHC) The soil's capacity to hold water that is available to the plant. This value is between the field capacity (maximum water threshold) and wilting point (minimum water threshold before the plant dies).

Availability of nutrients An empirical concept that divides plant nutrients in soil into available and unavailable forms. In fact there is a complex relationship between the two. Most soil-testing measures notionally available nutrient pools.

Blocky Cube-like or polyhedron soil shape. Soil has to close to flat or slightly rounded surfaces.

Bioaccumulation, phytoaccumulation The retention and storage of a particular contaminant or toxin within the plant.

Bioavailability The amount that an element or compound is accessible (or available) to the plant. In some instances, a nutrient or element (or toxic compound) may be present in a soil but is in a 'locked up' state that a plant root may not be able to absorb or utilise. Bioavailability is influenced by the plant species and soil ecosystem (microbial and mycorrhizal characteristics of the soil, pH, cation exchange capacity, soil organic matter, texture, porosity (which determines the amount of air in the soil), moisture, other chemicals present and dissolvability).

Bioremediation A general term referring to the natural processes, assisted by soil organisms, and live and dead plants (through biological systems with bacteria, microbes, fungi, algae and other materials), and by other natural processes such as oxygen and sunlight, to remove or break down contaminants in soil or water. This general term can be used with regard to contaminated air, water, soil and other matter.

Break-out zone A landscape area with suitable soil and in relatively close proximity to the subject tree, and with adequate passage through soil for the trees roots to grow through and be allowed to take up soil water and available soil nutrients. The 'break-out zone' is predominantly turf, permeable open space or adjacent garden beds. The purpose is to provide further soil resources to the tree for health, growth, stability and longevity. Some additional stability may also benefit the growing conditions of the tree (*see also* shared root systems).

Bulk density The weight of a given volume of intact or repacked soil. Also called soil bulk density. The measurement includes the pore spaces or voids containing air and is calculated in the dry condition. This is a useful measurement to determine the degree of soil compaction.

Calcareous Refers to soils that contain calcite ($CaCO_3$), dolomite ($CaMg(CO_3)_2$) or both. These can be destroyed only by acids, releasing carbon dioxide and Ca and Mg ions. However, sufficient acid must be added to completely destroy all the carbonate before the pH will start to decline. This delay is responsible for the soil's buffering capacity.

Capillary fringe The subsurface soil zone immediately above the watertable where water is drawn into the interstitial soil areas through capillary movement or surface tension.

Capillary movement Movement of groundwater into interstitial spaces through surface tension and attraction of water molecules.

Carbon:nitrogen ratio The ratio of total organic carbon to nitrogen in a soil or compost used as a predictor of nitrogen availability. High C:N ratio above 25:1 is usually associated with nitrogen depletion or nitrogen draw-down and low C:N ratio with N release or mineralisation. In compost, low C:N ratio is often associated with odour and ammonification. The ideal ratio in stable conditions is 15–25:1.

Cation exchange capacity (CEC) The capacity of soil to hold nutrients for plant use. The higher the CEC, the higher the fertility of the soil. A low CEC soil will mean the nutrients are easily leached out and hence not available for the plant.

Changing pH Soil acidity or alkalinity affects the ability of nutrients to travel from the soil to plant roots. Nutrients can only travel while soluble and are restricted or 'locked up' when outside their preferred pH range. To change pH, the following ameliorants are applied: to raise pH – lime or dolomite; to lower pH or acidify – sulphur or sulphate of iron.

Chernozem soil profile A very fertile, deeply rich agricultural soil containing a high decomposed humic content.

Chlorosis A pathological nutritional condition where the leaf veins remain green but tissue between the veins goes yellow. Usually associated with trace metal (Fe, Mn, Zn and Cu) deficiency or toxicity. Typically expressed in the younger foliage.

Clay loams Soils consisting of sand, silt and clay where there is a close to even proportion of each. Feels clayey and has a good cohesion. Forms ribbon 38–50 mm in length.

Clays Very fine-textured soils, with particles less than 0.002 mm. Form hard lumps when dry and are extremely sticky when wet. Heavy clays form ribbons 75 mm in length.

Columnar A soil type where peds are column shaped with flat or rounded faces. This type of soil can be dense and difficult to penetrate. Associated with sodic or swelling clays soils.

Composted mulch An organic material such as bark or compost that is used to insulate soil. The purpose is for improved soil fertility, nutrient and moisture retention and overall enrichment. The composted mulch should demonstrate compliance with the chemical and physical requirements specified in AS 4454 (2012) for soil conditioners.

Composted soil conditioner (compost) Any composted organic product that is suitable for adding to soil and that demonstrates compliance with the chemical and physical requirements specified in AS 4454 (2012) for soil conditioners.

Composting (although not applicable for installing landscapes) Composting is necessary for mixed vegetative waste created on site and the required temperatures needed to avoid spread of most weeds, diseases and pathogens. Generally this is considered acceptable between 55 and 65°C coupled with aeration, watering and mixing. All commercially obtained organics must be composted, except single source timber mulches.

Conglomerate A coarse-grained sedimentary rock form.

Controlled-release fertilisers Fertilisers that are encapsulated, which allows for slow release over a specific amount of time (e.g. 6 months). Higher temperature will increase the rate of release.

Crumb Porous, spheroidal peds that are can be easily broken down into smaller units. Ideal for root growth, water and oxygen permeability.

Cultivate Preparing soil for planting through the use of a hand/rotary hoe to break up the soil to promote improved porosity and aeration.

Depositional Erosional deposits of soil through wind or water.

Dispersion The separation that occurs primarily on sodic soils whereby the clay particles are displaced from the soil aggregate when the soil becomes wet. Gypsum is used as an ameliorant to improve soil dispersion.

Duplex A soil profile form whereby there is a sharp contrast between the A and B horizons, usually transitioning from a sandy loam to a clay texture.

Edaphic Derived from, produced by or influenced by the soil. Of vegetation: distributed according to soil conditions.

Ericaceous plants Heath plants in the Ericacea family (azalias, heather) that prefer acid soils and low soil fertility, and, in particular, low phosphorus levels. An acid soil and low P option specification need to be chosen for these plants. *See also* phosphorus sensitivity. Other plants preferring acid soils are in the Theaceae (camelias), Rutaceae (murraya) and Rubiaceae (gardenia).

Erosion and erosion control Soil erosion occurs through the elements such as water and wind, causing breakdown and loss of topsoil. Ensuring there is good soil structure by adding gypsum can counteract erosion.

Field capacity moisture content Arbitrarily, the amount of water remaining after a saturated soil material has drained against gravity, nominally for 48 hours. Field capacity is meaningful only for *in situ* soils and varies with soil texture.

Field texture assessment test A test used to determine the behaviour of a soil through moistening and moulding a small handful of soil into a ball or bolus. The texture class is based on the texture, cohesion and manipulation properties of the bolus.

Gap-graded soils Soils that have a range of particle sizes missing in the size spectrum somewhere between large and small (i.e. there is a 'gap' in the gradient). Also called 'poorly graded', these soils show better porosity then 'well-graded' or uniformly graded soils, which pack down hard leaving inadequate pore space for plant roots.

Geological terms *See* alluvium, conglomerate, depositional, igneous, metamorphic, parent material, residual, sand, sandstone, sedimentary, shale and silt.

Granular A way to describe soil that has a crumb-like structure. It can be gravel, sand or silt, but typically with little to no clay content and typically the soil is not cohesive.

Green manure A method to fertilise or nourish soil by first growing beneficial plants that are then turned back into the soil for the purposes of returning nutrition and organic matter to the soil and improving soil texture. Typically a legume crop is used.

Green waste compost Compost that is derived from biodegradable green waste that is high in nitrogen and therefore has a low C:N ratio.

Groundwater Water held in saturated soil and rocks that may be permanent or temporary. Only shallow groundwater can be accessed by most plants.

Hold point A term used in construction specification documents meaning that the contractors work must not proceed until each 'hold point' is satisfied to the terms of that contract and hold point.

Horizon A soil layer that is different to the soil layer below it (and above if it isn't the top layer) in organic content, texture, structure and colour which changes with influences of depth, change in soil chemistry conditions, age, and compaction or settlement. Fundamentally, each layer (sometimes called profile) can be divided up into five major horizons: O organic matter; A topsoil; B subsoil; C often clay; and R regolith.

Hydraulic conductivity (HC) The rate at which water is conducted through pore spaces or fractures in soil. Hydraulic conductivity, or K, is measured in cm/h.

Hydrophobic soil A soil surface that repels or is unable to mix with water.

Hydromorphic A soil profile that is developed through the presence of excess water.

Hyperaccumulation is a concentration of a toxin. Plants can actively take up and store within the plant body, and sometimes in particular parts of the plant (such as in the seeds of sunflowers and roots of marigold).

Hypoxic Refers to a soil environment depleted of some or all its oxygen, usually as a result of the respiration of roots and micro-organisms.

Igneous A rock type with volcanic origin solidified from lava or magma.

Infiltration rate The rate of which water percolates through the pores in a soil. Also called saturated hydraulic conductivity (*see also* permeability).

Inorganic sharps Hard visible contaminants likely to cause stick injury such as hypodermic needles, glass shards and sharp metal, such as can be found in dumped or landfill sites.

Inverted soil profile Occurs when the natural order of soil horizons, where coarse material always overlays finer, is inverted creating perching of water in the fine layer overlying the coarser layer because water is more firmly attracted to the fine layer than the coarser one below.

Iron/manganese nodules Hard nodules in a soil, usually platy or rounded, that may be composed of iron (reddish coloured) and/or manganese (black coloured) oxides. They indicate a soil horizon subject to periodic moisture oscillations and associated hypoxic and oxic conditions (i.e. periodic waterlogging due to impeded subsoil drainage).

Krasnozems Deep red well-structured clay soils of high-rainfall volcanic plateaus of Australia's eastern seaboard and occurring in other countries such as Russia. Also classed as Ferrosols.

Landscape soil A soil used for amenity and landscape purposes and usually affected by anthropic processes. The soil profile may be either modified from a natural soil or manufactured and installed using artificial components, intended to support a vegetation type chosen for the purposes of landscape design or restoration.

Large particles Soil particles, natural or foreign, > 2 mm in size.

Lateral water movement Horizontal water movement due to vertical flow being impeded.

Leaching Vertical loss of water-soluble nutrients through rain or water treatment such as irrigation.

Loam Consists of sand, silt and clay where it feels like there is an even proportion of each. Exhibits a spongy, smooth feel with greasiness if organic matter is present. Will form ribbons 25 mm in length.

Macronutrients Nutrients taken up in plants in large amounts. The major macronutrients are N, P, K, Ca, Mg and S, and are essential for plant growth.

Macropore/micropore Large and small spaces between individual soil particles that allow for aeration and water pathways.

Massive Grains of soil that are not bound to one another in an aggregate (i.e. the soil has no structure). Sand is compiled of single grain particles (*see also* single grain, apedal).

Mass planting Listed in title of a specification – grouped plants (often of like species) planted together in one garden bed, often to read visually as one landscape character or type. Typically used as large areas or swathes such as along roadways.

Metamorphic Rock type that has metamorphosed or transformed through heat, pressure or chemicals.

Microbial activity Microorganism activity that aids in the decomposition of organic matter, resulting in available nutrients for plant uptake and chemical and physical changes in the soil.

Microfauna Microscopic animals that are not visible to the naked eye. These include fungi and bacteria (*see also* microbial activity).

Microflora Microscopic plants, usually algae, that are not visible to the naked eye.

Micronutrients Essential trace elements that are taken up by plants only in small amounts.

Mushroom compost The spent substrate from mushroom cultivation, typically very high in organic matter, nitrogen, phosphorus, calcium and sulphur.

Mycorrhizal associations Symbiotic associations between plants and fungus whereby both mutually benefit from each other. The plant supplies carbon to the fungi and the fungi provides inorganic nutrients to the plants. The majority of higher plants depend on this relationship and it would be detrimental to the plant's survival without this association.

N:P:K ratio The ratio between the essential nutrients nitrogen, phosphorus and potassium. Nitrogen is important for photosynthesis; phosphorus is key for root growth and nutrient and water transport; and potassium aids in growth and reproduction. Fertiliser options range in their NPK ratios to cater for different nutrient requirements or deficiencies.

Necrosis The death of plant tissue due to foliar injury or disease.

Nitrification The process of ammonium been converted to nitrite by nitrogen-fixing bacteria to nitrate. Plants take up both ammonium and nitrate if the soil conditions are aerobic, but all ammonium will eventually be converted.

Nitrogen draw-down Nitrogen consumption by soil organisms in a high-energy (high carbon) environment. It can occur in soils after high carbon residues have been added (e.g. woody mulch). A test, nitrogen draw-down index, is used to measure the severity of draw-down.

Nutrient deficiency Insufficient supply of a nutrient that can hinder optimum plant growth.

Nutrient toxicity Excess supply and uptake of nutrients into a plant that may be detrimental to plant growth.

On slab Soil or growing media or landscape that is placed above or disconnected from the earth, such as in rooftop planters, container pots or elevated landscape zones, or in a landscape zone that has a membrane or artificial material such as concrete or plastic liner between the growing soil and the deep soil or natural earth or ground below. Also referred to as 'on podium'.

Organic mulch A mulch applied to the surface of a soil that is composed of organic particles, such as wood chips, pine bark. Benefits of organic mulch include weed suppression, reduction in evapotranspiration, improvement in infiltration of water, thermal insulation, promotion of microbial and fungal activity, reduction in temperature fluctuations to soil, and aiding in the reduction of hydrophobia/hydrophobic conditions.

Organic soil A soil with 15–25% organic matter content used for very high fertility garden and display beds often of a temporary or semi-permanent nature, such as annual flowering displays or planter beds that are regularly topped up. They bring benefits of enhanced fertility, high water-holding capacity and rapid root development, but are prone to oxygen depletion if they become too wet and to shrinkage as the organic matter decays. They should not be used below 300 mm from the surface. It will generally not be possible to make a low phosphorus soil with such high organic matter content.

Osmosis Movement of a solvent through a semi-permeable membrane from an area of high concentration to that of a low concentration in an attempt to create an equal or isotonic state.

Oxidation A process resulting in the loss of electrons. The most common oxidant is oxygen from the atmosphere.

Pallid layer A layer of pale or even whitish usually fine sand and silt between the surface topsoil and a clay B horizon. It is a layer of intense weathering and depletion of iron due to periodic waterlogging.

Particle size analysis A test method that separates and measures the proportions of sand, silt and clay in a soil. The object of particle size analysis is to group these particles into separate ranges of sizes and so determine the relative proportion by weight of each size range.

Parent material The originating material from which soil is derived.

Pathogen A fungus, bacterium, virus and other organism that can cause disease in a host plant.

Ped/pedality An aggregation of soil particles into variously distinct clumps. Peds can be seen when a soil sod is gently broken open to expose the natural fissures, gaps and faults and gently broken up to separate individual peds. There are different shapes or types of peds and these differences can be used to characterise soils (*see also* soil structure).

Perched watertable Groundwater located in the unsaturated zone and above the watertable (*see also* inverted soil profile).

Permeability The ability of water to move through soil (permeate it). It depends on the physical and chemical properties of the soil, notably particle size distribution (the range of particle sizes present), pore space, pore size and the continuity of the spaces.

Permeability class An estimate soil permeability range based on soil texture and structure. The classes we use are based on those of Charman and Murphy (1991) and are:

Permeability class 1 Very slow < 2.5 mm/h
Permeability class 2 Slow 2.5–5 mm/h
Permeability class 3 Moderate 5–20 mm/h
Permeability class 4 Moderate to rapid 20–60 mm/h
Permeability class 5 Rapid 60–120 mm/h

Permeability class 6 Very rapid > 120 mm/h.

Phosphorus sensitivity Species of plants that show adverse effects related to iron and other trace element deficiency if grown in soil with excessive phosphorus levels.

Physical properties The physical characteristics of soil are made up of several properties: permeability (the rate at which water travels) and water-holding capacity, porosity, texture and compaction. All these factors alter the soil quality and have a direct consequence on plants.

Phytodegradation, rhizodegradation, biodegradation: the ability for a plant, soil, or soil ecosystem to break down compounds. This process is influenced by light, water, oxygen, temperature and soil microbes and the complex interactions of living organisms including plant roots and soil.

Phytoremediation the general process of removing pollutants and toxins in soil (or water) through the use of specific plants predominantly but aided by living soil processes (such as bacteria, microbes, fungi, algae, oxygen and sunlight and other processes). The plants will either uptake the toxin and store it in the plant (bioaccumulate), decompose, degrade, or metabolise or stabilise the toxin or pollutant typically using enzymes produced by specific plant species.

Phytostabilisation is when toxic compounds are taken up and stabilised or bound together by plant bonds (which can then be removed from the plant or environment more efficiently).

Phytotoxicity This is caused by a substance that is toxic to plant tissue. Toxins can range from herbicides to heavy metals to high dosages of nutrients or highly acidic or alkaline conditions.

Platy An individual ped that is flattened and shows layers. It is oriented horizontally, which can impede vertical water flow.

Polyhedral Many sided, more than six flat faces.

Prismatic Similar to columnar structure and forms a pillar shape. The soil has flat faces and is associated with sodic soils.

Profile form Horizon layers deposited over time to create topsoil and subsoil (*see also* horizon).

Recarbonisation The process of improving the soil's function to sequester carbon and maintain it, thus becoming a carbon-rich soil, also with improved fertility and reduced potential for erosion compared with degraded soils. Carbon-rich soils and the act of recarbonisation of degraded soils assist in reducing greenhouse gas emissions.

Residual Remaining soil particles that are non-soluble.

Rhizofiltration The filtering contaminated water through soil and plants via plant root stores. Also called phytofiltration.

Rip, ripper The process of lifting soil using a tine to improve soil density, aeration and porosity, and reduce compaction.

Root plate The typical shape of a tree root zone grown on soils with hard or impermeable clay subsoils where 90% of the root system is present in the surface 300 mm and roots extend widely rather than deeply.

Saline soil Soils high in soluble salt content, usually sodium, magnesium chlorides and sulphates. Plant tolerance to salt levels vary, but the most salt-sensitive plants start to show yield decline at an ECEC (exchange capacity of the saturated extract) at 3.5 dS/m and the most salt tolerant plants around 15 dS/m.

Sand Material derived from rock and mineral particles. There is a wide range of sands, depending on their origin, but they primarily are derived from silica or calcium carbonate.

Sands Loose and single-grained soils, with no aggregate mineral particles. Unable to be moulded to form a bolus.

Sandstone A rock primarily comprised of very nutrient-poor, coarse silica sands.

Sandy loams Soils consisting of sand, silt and clay. The bolus is sandy to touch and grains can be seen, but silt and clay content is sufficient to give coherence. Will form ribbons of 13–25 mm in length.

Scarify To break up the organic and surface layer of the soil to expose some mineral soil and cause some degree of incorporation of organic matter. This is done to bury recently spread seeds, break up heavy pasture sod and allow the seeds of other species to successfully germinate. It is also used to smooth a rough-ploughed surface or to key subsoil before applying topsoil.

Sclerophyllic Refers to hard-leaved plants designed for reducing transpiration. Most commonly found in dry hot areas. Common plant families include members of the Proteaceae, acacias, tea-trees and eucalypts.

Sedimentary Refers to rock that is formed from the deposition of sediments.

Shale Fine, silty clay deposits that forms stratified sedimentary rock.

Shared root systems (after Sjolfeld 2009) Shared root systems can occur where trees of the same species or similar species can (but not always) fuse (or conjoin) feeder roots together to enable shared soil resources (water and nutrients). Shared root systems can occur naturally in forest systems and can enable greater tolerance of a stand to stability, climatic and growing condition extremes (such as drought, wind or shallow soils). The dimensions of soil required for shared root systems to develop depend on adequate provision of 'the eight key factors' (**Appendix C**). A shared root system zone needs to be wide and deep enough to provide suitable growing conditions for tree roots to establish conjoined roots and trees to be spaced appropriately to promote conjoining of roots.

Silt Particles originating from silica or feldspars.

Single grain Grains of soil that are not bound to one another in an aggregate (i.e. the soil has no structure). Sand is compiled of single grain particles (*see also* massive).

Site-won/won Refers to soil derived from the site.

Slake Soil aggregates that breakdown into smaller sized aggregates if exposed to water.

Slumpage Loss of soil stability or sinking of soil. Generally occurs on steep gradients or where vegetation is disturbed and gravitational forces cause soil to move downwards.

Slumping The reduction in volume of soil over time through natural decomposition of organic matter and settling.

Sodic Where more than 7% of the cation exchange capacity is occupied by sodium ions. Such soils tend to disperse in water and show high erodibility.

Soil auger A hand tool used for boring profiles from soil.

Soil conditioner An additive incorporated into soil media to enhance the physical and chemical condition of the soil.

Soil landscape A grouping of soil types associated with a particular geological origin, landform and topography. Within the grouping, the association of soil types is uniform enough that the presence of a certain soil type in any given position can be predicted with some certainty.

Soil structure The shape and arrangement of soil peds (aggregates or structural units) in pedal soils.

Specifier The landscape architect or designer or person who is writing the specification report and deciding on the type of soil to use.

Spheroidal Refers to rounded peds that can be easily broken down. Peds that are very porous are termed crumb; those that are less porous are granular.

Subgrade The finished constructed surface upon which the soil profile is to be built up. Subgrade is usually either mixed fill or the cut subsurface of the original soil profile. It may be possible to ameliorate subgrade to form subsoil before placement of topsoil, but salinity, extreme sodicity, rock content, compaction or other hostile properties may mean subgrade cannot be ameliorated and subsoil must be imported.

Subsoil The layer below the topsoil that is imported or ameliorated from site subgrade conditions to form a suitable rooting anchorage medium, moisture and nutrient store. Subsoil is imported in completely reconstructed soil profiles, but may be produced by amelioration of the subgrade before placing topsoil where the subgrade is suitable.

Sulphide The reduced form of sulphur occurring in anaerobic environments. Sulphide soil typically shows the black colour of iron sulphide and characteristic rotten egg gas smell.

Sulphide oxidation *See* acid sulphate soils.

Surface tension An effect of the intermolecular forces that occur between the bonds of water molecules. This minimises the surface area.

Symbiotic and non-symbiotic bacteria Bacteria living in symbiosis or mutually beneficial relationships with their hosts, such as nitrogen-fixing bacteria (*see also* mycorrhizal associations). Non-symbiotic bacteria perform the same role (nitrogen fixation), but do not rely on a host for an energy resource.

Texture and texture class (after Northcote 1979) The relative proportions of sand, silt and clay in a soil. Texture class is assigned by a manual manipulation test on a moistened and worked 'bolus' and assigned a class. There are five major classes but these can be further refined using such prefixes as 'silty' – thus 'silty loam'. These classes are sands, sandy loams, loams, clay loams and clays (see the definitions of each).

Vermicast/vermicomposting The process of organic waste decomposition through earthworm digestion. The excreted waste from the earthworm is odourless and rich in nutrients and organic matter, so is ideal as a soil conditioner and fertiliser.

Unhumified O horizon or unhumified organic matter layer with leaves, twigs, etc. still obvious.

Visible contaminants Artificial or synthetic objects such glass, plastic and metal present in sufficient quantity as to cause adverse visual impact.

Water stress The result of the reduced availability of water to the plant during drought or from increased transpiration from external factors such as high temperatures and soil salinity.

Water-holding capacity (WHC) The minimum amount of water a soil can hold that is available to a plant against the force of gravity. The capacity of a soil varies depending on its composition. A sandy soil would have a lesser WHC than a clay soil due to the clay soil's ability to hold the water.

Index

Printed and bound by CPI Group (UK) Ltd, Croydon, CR0 4YY

13/03/2025

14640381-0001